Does Anything Work Shattered?

Does Anything Work Shattered?

The chronicles and thoughts from an unlikely source, an unlikely finish to what is a great controversy.

This was not intended to be as it is, it melted within our grasp

Michael Wolff

Copyright © 2015 by Michael Wolff.

ISBN: Softcover   978-1-5035-9934-5
          eBook   978-1-5035-9932-1

All rights reserved. No part of this book may be reproduced or transmitted in any form or by any means, electronic or mechanical, including photocopying, recording, or by any information storage and retrieval system, without permission in writing from the copyright owner.

Any people depicted in stock imagery provided by Thinkstock are models, and such images are being used for illustrative purposes only. Certain stock imagery © Thinkstock.

Print information available on the last page.

Rev. date: 09/28/2015

To order additional copies of this book, contact:
Xlibris
1-888-795-4274
www.Xlibris.com
Orders@Xlibris.com
717454

Through experience or time does it take to add to Nirvana?

It is not only who you know but what you know as well, then it is what experience there may be as to what time it may be provided that is most important to life.-Michael Wolff

Who in the entire world is too perfect to ask questions of desired knowledge, curiosity, or even faith?

To the order of the day when we attend church if we go at all, knowing that this may not be the only location with which its organizer had fallen in love with the page and not the sage, as my childhood grew into adulthood, I came to understand what, within my experiences, of which I could have to do and not so much have the desire to do. I desire not to do this based upon the reason Jesus Christ had been persecuted, enslaved for little more than a handful of silver, and not to forget the others in history attempting to do better for the greater good of the rest of us in the honest to goodness way. So as to tell you that I don't want to die for what I do, in all reality, if that be the case although possible, history could look to my accord as another statistic, though one yet does not exist to my knowledge of whom have been persecuted, stagnated, killed, or enslaved for trying to do the good things instead of what the flow of life may be, I will still have made my mark within the given history time is, if even one persons eyes be opened, or that time run its course out of time.

Not much exists in pleasure within life more than that of earning something you have set your own mind to. For instance, a vehicle, given or earned, which is more appreciated? From one individual to another, the first of these will never fully understand the other as well as the other way around. Contextually speaking, if you yourself are spiritual and believe in or have the meditation to look to a greater power through your own enlightenment, then a salute may be yours. If though you indeed only attend any sized congregation who worship a person for an almost singular point of what criticism may include, then no salute will be from me and I shall give my reasoning for such. First, permit me to clear things up through the definitions used and the reasons they are used against the ready and willing people, onto the way those same definitions could have been used and should still be.

Maybe possibly you believe as you're told through the fellowship of leadership and the congregate population around you. In any effect, the exact opposite may also be true as to Newton's law of action reaction. Balance if you will is supposedly the key term within either way to believe, spiritual or non spiritual, titled God or not, including popularity of whom you have been raised to believe in if that is in your nature. This first query into definition I presume is a belief of many around throughout time since the story of Christ had begun. Though a story these days, I have no possible proof that miracles had been made, though I know within the natural mind, they are quite possible. So possible in fact, that in the people and the mental adaptation of natural evolution handed to the next generation, without assistance from technology or drug, the availability to have the newborn child speak at birth is not out of the question as a possibility worldwide after thorough effort is completed, but should have been here by now considering the time it has been from Jesus' death into today if He, God's only son had completed his quest as desired, and not been of an untimely death by an external force. I myself do NOT consider Christ a failure though his works were a failure only after his message which had been to the people and their enlightenment of freedom and equality, was used as a guilt because if you ask yourself who around you would legitimately die for you and you alone, as it is portrayed and I'm sorry to say profited by, Jesus' Christianity is the largest misunderstanding I've ever tried to understand.

More correctly though, if not entirely Jesus Himself, but the works the book that had been written over sixty years after his death, came to be known that it were a recollection for He was not yet found though awaited for.

If to get closer to the root cause if not that of itself, this entire globalized profit driven brotherhood of so-called believers in Christianity were to unveil the truth of what has been done since the beginning, up to and not forgetting the Christian crusades of mass murder defined today as genocide by sheer numbers of the slaughtered innocent who apparently were by this Christian God given the freedom of choice, though no longer have as their own choice in sight, this truth would re-arrange the globe through the indecent exposure and mass population near panic. Not to worry though, I have this under control and full use of reason within my own basic human rights though they in themselves are no longer as basic as they may seem and shrinking every day though you may not directly see it done as so explained.

Ok so the portrayal of which is Jesus, that of Him Giving His life for your sins and all of mankind's, is the belief of apparent truth. In all reality, he had been enslaved, tortured, and killed against his will. I mention against his will for a couple of reasons. Reason number one; survival is one of mankind's strongest individual adaptations because the will to live is the will to survive, no matter how the survival is taken into account, it is still of the most basic instincts an individual cannot merely give away. So the subject of the matter that He had GIVEN His life is Non-Factual. Though I do understand the argument that He or His Father is the know all and see all, past, present, and future, I have to admit that that is also an overwhelming degree of control not unlike guilt though this guilt is smaller than that of His life for your sins. The sister to guilt is fear, and fear from an Almighty power when portrayed as all knowing, is a great motivator even though it is of the wrong movement to the supposed end goal of salvation through His Son. Unless that end goal of the "Church" is to actually have that of what it teaches, be the sole purpose of life's own defeat and demise. Those two basic things of guilt and fear are though very powerful, but addictive to have under your control, but in as much as that is, to have enlightenment for all of mankind and then only to have it die with you for a trade off for the forgiveness of your sins is fairly corrupt when you take into consideration the same adaptability this control has over the believer as the believer has his own adaptability for survival and this becomes a profitable account to live a livelihood of less than inadequate resources to those who are the listener and hand around a collection basket, full or not, is in such a self defeating explained way, it distracts the mind to a congregate purpose but to who's purpose is it to? The two most basic functions survival has created through necessity to have life continue thus flourish in theory is music, art, and poem. All three of these I understand to be different, but, it is not understood that they are all of the same system of survival and doing the same exact thing? Consequently all three are one in the same in fact. The sole purpose of such a thing is to get those in need, out of an unhappy rut and get by for the moment happily distracted. Adapting to what we have been lead to believe is looming over us and always ready to strike rapture through Christianity, and politics as well. This thing survival has permitted through necessity for the comfort life is and is not, is they joyful experience until the guilt comes into the elephant sized emphasis which has grown like vines into everyday lives and takes that moment or experienced joy away until the adaptation comes into play when the individual mind creates the reasoning to continue, or entirely assimilate to blindness and acceptability of what had been given as a so called belief, or truth, but a lie either way.

Reason number Two is a slightly different approach but plausible as well as reality; when someone has something great to accomplish and tries to do so, has the will to do so, either for self or an entire generation and then the next and so on, I would think that such determination and enlightenment wouldn't just GIVE that away for his mark on history to be but a twisted turn against those this individual had been trying to help if not free from stagnation, unfairness, and greed.

Onto today and the uses this message has been twisted. Assuming a fundraiser for appropriate so called needs (being a glamorous or actual necessity) is held and funds acquired, the overage isn't talked about when let's say a pathway to the doorway of salvation on a Sunday could look like. This overage may actually be the funding, or more important, is the accountability it becomes to those who feel as though it is more accepting, so they accept this beauty more as their own beauty for they see it and then associate it with their self, and within their sub-consciousness, guilt unknowingly and the two associations of guilt and beauty become one tangible controllable thing (a person/believer) even without the use of fear that can be used at will for many other reasons or agendas. Be those other agendas, wars which are also profitable, politics when in coercion with or next to as a similar control, or the infiltration of the individual spreading this twisted message as his belief MUST be true though he/she has not given it much thought. The Bible itself has many teachings to take guidance from; though it is not in a natural form of assistance overall to the good this era has within it any longer. Re-written numerous times, by numerous people, adding up to thousands of different interpretations and uses in thousands of ways, it's too big to be seen at a glance that it's the thing that is enslaving the individual mind. If for some reason you don't like this glance because you haven't bought into this corrupt society of belief, you are defined as an atheist which is unacceptable to the standard society has lowered their selves to be but blindly as they have done so, you as an atheist are not far off what the truth may be. The religion in its essence is too "In love with the page and not the Sage" they claim salvation through. Have your own opinion, follow the flock, take the lead for yourself or not, I don't care as much as you think I do…it's probably more than you'd imagine.

Now if for some reason you're still within these pages reading on for delight, splendor, stupidity, or even curiosity, I thank you for your consideration to yourself and the price you have saved yourself from paying. And that savings you now have is time which is entirely more splendid if you care to use it wisely than the currency we use *for* our time within today's impossibly upside down society.

Now at an impasse with success, that of the majority and their survival while not necessarily mine, if taken into account the phenomenon of Déjà-Vu and the possibility that the belief of success is the only answer, (and now closer to your grasp) then the impossibility of failure becomes true and at times evidently felt to the understanding of existence and the unity it has with all things you are, and all things around you. Your choice is who you become with this as it has been of the original message in my opinion, take it or leave it.

Here are some wise words from whom I will name when finished for it's a little obvious to some. And I quote "Because of the self-cherishing attitude we think we need to protect ourselves and those we call dear ones, out of this attitude we attack others, wanting to destroy them. This way we create suffering for others and anger for ourselves, not realizing it is our self-cherishing attitude, not others, the source of our suffering" from His Holiness the Dalia Lama-Lam Rim teachings 2013 Sere Jey. Greatness like this within the world is too few and far between enough for everyday lifestyles to be adapted to, which is why those who have done as much for teaching are those few who are commended so often. Either way this greatness deserves reward for spreading such depth into the mind becoming awe of onlookers, minus the difficulty that this quote had been through just to get to where it had.

Knowledge is accessible throughout the life you live, and few people within that life you live seek things from their own willingness to learn on their own, and not by force, although the force is there most times. Albert Einstein had integrated the largest imaginable forces within the cosmos and then worked down from there systematically to solve a problem he had deemed worthy of solving. Reportedly he didn't even want to memorize his own phone number because it would only be taking up valuable space and be distracting. However that may have been, stimulation is improper without honest knowledge in the areas desired. No barriers exist really, barriers are only distractions able to be set down with ease and disregarded for their complexity given the proper knowledge. Prosperity from what is, to those who can, prosper, when there is an adversity to their honest human rights as basic as they are to the largest factors of those same rights, are in accord to that of a responsibility to teach others as honest as a human right actually exists, not manmade laws which strip confidence to the bone. Self reliance and thus optional knowledge give life the opportunity to take over if you don't care to take care of it for yourself. That won't include peer pressure in essence of a fool or their journey, should you argue with a fool, you become one do you not? Unable to argue this next point against what it reveals, if one or more people do so try, then I may just entirely walk away if their argument is even remotely weak. The placement of many unrelated facts, situations, time, people, analogies, and metaphors placed together to bring together a topic of a given situation, is a difficult arrangement to have credibility with when life is taken for granted on accident and entirely planned by others far before their time, ultimately to correct the belief, if even that of which currently is a Savior represented incorrectly. Currently a faith not known to others alive is that I, Michael David Wolff, am here and here to fight in an unlikely way. I have that mentioned faith in myself. That previous statement has merit to why so many people have claimed God not real for He speaks nothing for such a long time for the unrealized and apparent chance to let the corrections become evident and in place, though they do not come from nothing and form to make the correction in reality abundant enough for actual use, though they have not and the reason why I am here fighting unconventionally as currently necessary though able through Him in other ways able to fight in many styles of form. So here is the apparently unconnected situation, and the possibility that this is the reason for silence, and which gave reason for an intervention, perceived as divine or not, humanity comes first. A utensil is best described by naming first the use it is intended to serve. By following this up with a general idea of its shape and size, we have well prepared the way for a comprehension of details. Now I ask you don't shoot the messenger, but isn't that in the same context and effect that the story of John the Baptist who lived in the wilderness and considered then to be wild and crazy for doing such, though well noted to have been the person who had paved the way for Jesus Christ's coming and the life he would live? I may have paraphrased there but I think that's the way I remember being told of John.

This fight has the given time for completion. What is a limit when if Jesus had risen from hell after three days in hell? If for some reason He hijacked the entire earth along with Himself then we are and have been immortal in the sense that He has the only reset button for completion to the effect of success (the end goal). In addition to that, what is the difference of time from there to here within life and even greater places like the heavens when time from heaven to Earth, life is in heaven three days to earths three hundred and fifty some years? Days of perceived time in Heaven to around three hundred years here on earth as told in a biblical story sometimes omitted from teachings to subordinates within the brotherhood that keeps teaching those who lead the wrong way as the plan had planned. This way, only the few and far between will figure the system a lie, though that in itself is not the full effect, for there is a God, and an afterlife, and miracles, but when in a belief beaten, the look forward is much harder if believed more faithfully. Such things are like the way trust is within family and betrayal comes about and it hurts more. Removing trust for later uses unless the falsehood of trust is believed once again. An evil cycle such as this is the fight I am against and those who teach such things as it is good though it is not, will have to either thumb the highway or go with the program and legitimately open their eyes to the undeniable realization that they have been taught wrong if they hadn't known already. For instance this billion dollar industry when taught wrong is productive enough to continue itself for no more than more power over those who they lead, if corrected into the humane universal ability to be and do good and great things, in the honestly good way God would approve of, the possibilities of the overabundant happiness and availabilities of equalities as a standard being raised instead of lowered over and over will become so apparent and swift, that it soon can be forgot, though remember that prosperity also deems responsibility alongside it. When successful for a period of time enough, there may be retribution enough for miracles to become seen by those deserving.

I am not a poison, nor am I an anti-Christ or even a blasphemer to what is right, I don't claim justly living or thoughts for I have made mistakes and have for a long time been as confused as a cactus. I merely have the questions to fill in the blank and, I do however have thoughts of who I may resemble and in actuality be, where I may be from, and as I have already announced that I'm here to fight, permit myself the resemblance to the recorded children who have supposedly recalled their earlier lives and in such detail that science is at a baffled state of reprove. Life should not be a fight such as it is, nor is life supposed to be this far from the reason life is created. Perhaps my imagination is leading me far from the reality known, yet I have felt this undeniable tug and at times jerks towards what I need to accomplish if not at least attempt. You're reading one of what could be many of my attempts.

Rumor has it that within the gates of Heaven, the question had been revealed as if there would be any of which heaven has within it, take unto their human form and fight and there was a small pause before reaction had started because the reason for this had also been revealed. God Himself would not destroy this place He had created but is so distraught of the wrongs within, gave option for intervention. The remaining Arch Angels had waited for the others to speak up and none had done so, they then had spoken amongst themselves in their own way and only one had emerged as to take on such a task for the Creator. Arch Angel Michael stood among the gates of Heaven and as from a cloud, jumped as one would from a plane and the regretful ones who had been silent regretted not answering as well, also tried to jump only to meet their faces to the bow for their action was too late and now had to stay. In thought as I had been growing up, I don't remember the exact age I had been, but I too feel like I had lived another lifetime in another era. As a child I made my mind up that I would figure out what I would have to for another triumph over the undeserving liar and the problems he has made. Now there is that word imagination again. I swear on my life I'm not as wrong about this, but as for what follows, take it or leave it, I feel it.

Ganked is the closest explanation that could be derived from the sound within its entirety from beginning to end as the first Arch Angel had been cast into hell by Arch Angel Michael's sword by God's orders and the help from Grace. Ganked sums it up with all the movements, moments, and the finished deed. The use this word has today in whatever way is laughable at best. I attempt seriousness at most times because this is not a life of distraction, I consider my life to be of a purpose and currently will have an outlook that it is "as if" it is impossible to fail.

Fundamental to my adaptation to a mental status of what some believe to be multiple excuses to a simple diagnosis or many of them, the world in which I call home for an unknown period of time, is the acceptable possibility that the Arch Angel Michael is in fact real or not and or imaginary and considered by an atheist, nonexistent stories told to children for it to lead them for no good reason. At particular intervals throughout my life, as my knowledge gains within me, I must put two and two together. My full name is Michael David Wolff Birth date 6-26-1986. Given personal security reasons my social security number will be excluded from this works of mine though it too has signifying relation to my thesis of who I actually may be. As records show, in 79 A.D. Mt. Vesuvius erupted and buried Pompeii. Now merely two thousand years later, there through S.E.T.I was the WOW signal found and as it is reportedly longer if you consider it could have been moving at great speeds, there within this signal may have been a message of binary code and to the biological number a human being holds as DNA. Don't get me wrong, I'm still a little confused if this be the case, but if a human being can be sent through space in a code to another planet entirely different from their origin, then our science fiction is not that far off and though I was born eight years later, you must account for the time it takes for a human fetus to be born, and not mentioning the placement it had chosen. First my name and it's significances to being that of Arch Angel Michael. Starting with any of the three names in any order, I will start with my middle name for my own reasoning too tedious for explanation as to why. David is my middle name as I was given this to me a little after my birth. Going back to the first important person named David in history is the man who King Solomon had, and who had written much if not all of the book of Psalms and quite possibly more than we the majority of the congregation has, or will ever see. Whether that be true or not is unimportant, what is important though is that it in itself is an important aspect to life and human behavior to live by, being of such wisdom in many forms still relevant to today. This book within its experienced perception of mine has repeated itself through different ways, saying the same things at times using another word compilation regarding a few different things. If taken into account the idea that a signal is able to transport a being, then this resource of repetition may in fact be a code for the human genome and the uses it has through its own code, to send the code of a thing that could be greater than technology even allows itself to admit, science fiction to imply or even stretch itself to, and what science will do now more than investigate. Reportedly written through poem, song, and lyrical means, the repetition of the things within it are attempts to cement the good that is within every man woman and child if given the right direction(as it had to be hidden in plain sight such as Nostradamus had to do). Whether or not the book of psalms was merely written for itself to be alone and later added to the Bible or not is unknown to myself, (though I should find out for sure) it within itself or added to the Bible later is a very powerful tool for life's exploration. So to the root cause of where I like to look when I look at my middle name and find the importance, I look to this man from my history and thank him.

My last name, as that of an animal and not just any animal, but one that can be vicious, grow to great sizes, and to some cultures either be sacred animals or a cursed one for its varying degrees of temper. There will be more on my last name later on why it is what it is further within these pages. Back to the animal, alone or not, the wolf is a dangerous animal or beast if you will. The possibility that all of mankind's best friends, companions or working animals being canine's coming from this one animal is not an unlikely possibility. I will make no apology to Darwinism relating ape to man because it's such an obvious distraction that I won't even address it further. The name that had been placed after my first, David coming into the picture from where my eternal self had come from, the animal had been created first therefore; David should be mentioned secondly within my explanation as well as my given name, not just a thing of fate. In addition to man's best friend and its analogy, continuing to the possible viciousness of its original instinct, the look coming into view through the latter of my current years to myself and what has been greatly used is that I resemble facially, and quite possibly more than that in a physical aspect, Jesus Christ. Bear with me if you will the imagination for fiction to become fact like science has done so fairly often to make its own valid point over and over though many times wrong and disproven only for it sometimes to be re-proven. I may in fact only be daydreaming but I like to tempt my own imagination and stress the places it can lead to in thought if not life itself and its direction it will take me. Also I must ask that you continue bearing with me that the possibility, the idea that Jesus' message, of which is no longer as what it was, (actually used against the populace) but the vicious way He is foretold in his oncoming return only in the days of Judgment for that of the entire human race on the very last day, is only the form of control in its longest use. This is a form of control through guilt and the brother of guilt is fear, also a form of control if understood in its primal uses and not even as precise as the primal size of human nature. If you look at recent activity, daily happenings around the world, and recent historical events, things are changing direction from the calm and cool to the oh shit, lets run away. As things like this become more frequent, the "End Times" are reported to have the next day that of the end in effect an everyday turn of events, more and more common so it is as the acceptance of another genocide and psychologically for those who live to tell of it, more easily digested if done so by the works or disabling works of man. That's why I fear if I do nothing to change what may be coming or not, it will undoubtedly become a self evident and factual thing. Continuing through that imagery be it true or not, all of the previous information is presumably true or not though little in comparison to what has been a warning (many types of other warnings as well in fact, quite possibly through mathematics of what births there are in any given day and or era to the addition of the next and so on, the plan as to have the narcissism of a hero or an evil that is to be((the bringer of death)) to our era and onward from its own place) for security and thoughts of safety or defeat, was to one person's name in particular for today's time and that name is only of which I know through my ears and minimal education, the name of Barak. The Bible mentions him, the notorious Nostradamus has many things which may be relevant to today and has a respect of the way he has been correct to some degree of the belief given. He (Barak) in some way from the era mentioned, to the

Does Anything Work Shattered? | 11

time of today, became a thoughtless minimal threat until the look back on how he had or has conducted himself while in this nation's most influential position, in the same way Nostradamus is repeatedly arguably wrong in the sense that his quatrains are too coded to understand. The warning before he took office was minimally mentioned to many who had known or not, and therefore the warning plan had been useless in most of its effect if not flawless by those who had set this into motion. If a mathematical conspiracy theorist not using math anymore but instead spirituality, adds up things his/her own way, I Michael, who looks like Jesus, could in fact be the Arch Angel, could in fact also be in His sons image here to fight for mankind's success for reasons that will by God become clear, though the only Angel from heaven to come and battle, here on God's behalf to battle or make a decision to stop right here and conform to what the plan of what has been for decades, or fight as Christ would for intervention and the successful future of mankind and his survival, or as the definition of the return of Him(liked or not liked) for the humanity and reasons God has been told to us as having mercy, or merely a man who is chemically imbalanced and medically "unstable" at times. Physically I bleed like all others, and to that effect, can and will come to a point of death. On the other side of the scale I can be seen from is that of an Anti-Christ here to dissuade you from the religion in all its effectiveness. How would you decide if you're not allowed to make your own judgment for belief yourself may have? That is one of those taught things which is wrong to teach and always will be.

Anything further from here on, within these works, the thoughts or inventive thoughts of a product I may use or create may be used by myself as imaginative theory with or without cause for basis of clarity and, notwithstanding the ability and or desire or need to prove these thoughts and theories correct, though they are mine and mine to give, and as if it is a gift to the entire world, no one single person shall be able to patent such things as it is to everyone, a gift. Extremely advancing technology at minimal cost created a boon to what could be next, and my gift is that of what is also super advancement to everyday life and purchasing power.

Though not as He (Jesus) is portrayed in paintings to the full physical extent, I feel it necessary to tell you I have on more than occasion believed that I am in some way, an extension to and from Heaven no matter who I may actually be, but my resemblance to Jesus is the difference as the difference mankind has adapted to, for their natural nurtured survival from the time of His life to today as in their physiology and thus my facial and body differences and or similarities if you will. To some definitions I could be called blasphemous, even though looking like someone is at my own opinion, a liar and deceiver, or even and not disregarding all the way to the Anti-Christ by those who don't like difference being the thing of that in which will change their lives and inner workings to their livelihoods. I upon birth had been born with apparently one operational kidney and two genital hernias. Therefore the laws of man which had kept me from entering into the military were utilized by higher powers, even though I had tried every American service available though none would take me for my disadvantage of only having one kidney. Their look at this is if I were to be injured and then on dialysis for the rest of my life or having an implant then anti-rejection drugs is too costly a thing and therefore avoided. Would that not be the same as if I were to ACTUALLY be Him? If to live in His image, I'm doing no wrong. If the kingdom of heaven is within all of us, well then, as I look at this situation, I classify myself and the fight I currently am in called survival, want to share with my audience the will of the imagination I have and thus the majority of this book as such.

I'm told we are all on the same mountain trying for the top or happily sending down pathways for others to reach our position more easily. Even if that's not the case and entirely opposite, then this mountain is also survival and a network of connectives that not so many choose to look at, or into for their own belief of thoroughly intentional conscious actionable thought, thus they are choice for misguidance, and choosing it FOR themselves. Internally my thoughts to another are mumbled in a crowd, but to another from my perspective are either blank, destructive, indifferent, uneducated, overruled, over simplified, lazy, crazy, and many more that could be mentioned. Here, I can give you a piece of my ideas for a moment. If for instance the original message Jesus was spreading were to be as I have said it has been, though oversimplified and used as a life to martyrdom, and not his message at all, leading the listener to guilt and fear, denial, and blindness, then in fact the Christ you believe in is only a figure of temperament that controls you, wars, politics at times, and great many things alike and not alike. In that case, you have a false idol and worship a false God and in effect will behave as you do and have done. Just to survey the place He created, only to return to the definition as that of being the anti-Christ in the return of the King (Another of mankind's destructive creations), is destructive to a many fold way of uses against the lead congregation, and is supposed to come as an anti Christ item of wrath deemed necessary for "The Greater Good". Hold on to your shirts everybody I'm not purposely making any attempt yet to change your life. My indecision as to whom you may or may not worship is not my choice and

I willingly admit that, but if you look at the powers that be from not only His time but all the way up to our time, then not to forget that Christianity had made genocide a plausible thing because of the crusades (even though it had been done before then, but not to that extreme) the same exact crusades in which had made Christianity the force of man and not God because the thing God had given into the garden as choice, thus mans given choice as to whom they could worship after eating the forbidden, this whole ordeal seems to me more than a little off and misleading in and of itself as it is. I do not ask nor do I want anyone to follow me as their leader because that WOULD be narcissism, all I desire is for you to see for yourself what you want and get yourself there as easily and honestly as possible. I'm not against Christ and or His message, but to the powers that are in power and of whom have been in power ever since His unneeded death. The message He had was turned into a position of power to control and lead a life, and that life, not just that life of their own but the "flocks" as well, having a life of free living before the congregations own eyes and making all things the church has done and plan to do, far in advance or even tomorrow the things of a justly cause and necessity, the profit they receive is like the drug no one has mentioned in awhile. The limelight is an addiction to the weak, as well as those who are perceived as not weak, has great influences throughout the world within the web and circulation of media, as well as grown within this generation not mentioning the previous ones, leading into the next many. Not only are the weak those in the limelight, but those who are less likely to be blamed or shamed by judgment who follow this false idol lifestyle, poisoned by the idea that they too are living the life of luxury, because if you're taught wrong, should you be punished as severely as those who taught you wrong on purpose?. Very similar to the church wouldn't you say? Politics use the form of control as termed "For the Greater Good" though almost entirely through fear. Loss is a fear, not having Maslow's basic hierarchy of needs met is a fear, and if you have the limelight as one of many possible addictions, then the luxuries you have no longer within reach is a fear, forced taxes are a fear from not having enough to losing everything for not paying tax, and many other things that are deemed for the supposed greater good are in fact as corrupt as if the devil himself had planned it. Politics had better behave. There is more work to do there than any other area combined if you look at the entirety of the situation that it on a daily basis crumbles into the lives that they say they lead for the so called "greater good" and only living that lifestyle of power and limelight, freely taking and giving less than taken repeatedly through their far too large brotherhood of liars and cheats, murdering innocent people around the globe for their own ego to be temporarily quiet and not hungry. For instance the railroad industry had not made dime one after the government involvement because they paid by the mile and the cheated system taking a shortcut and not doing the necessary work for it to be done thoroughly enough for profit, and out went the window the Bible and the saying "In God We Trust" when the Bible says to do something is to do it well and the window was the opportunity to prosper for the long run of social status, not the immediate gratification of their own lining of a pocket. Then there is this one guy who had made the great northern expansion on his own dime and only made the next section of track when his profits began to offer another advance on into the west. The government had tried multiple times to lay laws before him and heavily tax his ethical business behavior but he triumphed enough to reach the west and desired to create more healthy business and start trading with the Orient. The first shipment was only a trial to see if they liked our wheat enough to continue trading from that of a freely given gift, if they desired more, for the uses were many, the trade would be made and guess what, it worked beautifully. Nothing and nobody cheated but the greedy government until the trading laws had been draught up to have it become impossible to make profit for James J. Hill known as Hills Follie finally beaten into submission after winning over the government for nearly twenty some years or more. When looking at how the political ways of the railroads were subsidized into the cheapest bidder, and when the tracks weren't a uniformly made size, oh well the government knew that would happen and probably underhanded the table cloth to have it done, made sure to have it redone, and then when people started to complain of the prices to ship their goods, the big government decided to step in and "investigate" the problem and started pointing fingers at the companies claiming they were the problem and were wasting the time and money of the taxpayers when all from the beginning the whole scheme was to finalize all of this to be able to have that limelight, the pocket lined lifestyle, and the power to do so at their will because of the investors and interests paid by those who had gotten fraudulent smiles and handshakes in the iron age.

Will somebody please take a moment to click off the light and then back on the same instant it is noticed? This is today, this is yesterday, it had been the same in the crusades. Wars and wars only for profit and power, glory and fame, genocide for the future of control…what do you think is next when there is plenty to eat like there is today but there is no one who has enough to purchase what there is? That's not a happy meal, that's more than a cloudy day. Storms brew like that of which was hurricane Sandy, and believe it or not, I knew would be because of something called 'wing dams' off of the west coast which are only large manmade dirt piles that reach the surface from under the ocean only for the desire to have more warm water find its way farther north. If you mess with the oceanic current in a major way, creating a shift even if thought to be a small one, later down the line, the after effect is a very dangerous and costly thing to keep control of, because you cannot control the nature of things not within your control. This control had not been given unto man as the animals and living things had been. Then without their prior knowledge, the Japan nuclear power plant exploded, and had not been properly cleaned up but covered up for the ongoing damage increasing the filth that oceans life would have to endure until it reaches the next and so on. A floating garbage collection reported seen from space. Hemp had been grown within Chernobyl and had a recognizable reduction in radioactivity in soil...place that in large swaths in the areas of infection within the ocean and large enough to be called a plausible attempt in multiple areas of that plausible attempt, it acts as a cleaning system...sinking the gathered plants in the deep until it is safe for global uses because there are far too many to not use this as a renewable resource, not to mention the TAX available, because this plant can grow in extremes of their environment, and many times the rate of traditionally used things like paper, and the first of many great changes can begin to reshape the broken days that are here now, not mentioning the changes that have gone into effect (a super computer) that are actually almost unrecognizable though incredible that they are in everyone's pockets. It won't change what ANY politician says to me, I don't like the ideas they have or the ways they fulfill their agendas because they're not using honesty and clarity. It's their organized theft and destruction as the brotherhood and sibling to the false portrayal a higher power has been said to be. I'm now going to make a reference to a book. Oddly yes I believe self education today is more plausible than that of which the government has given me in public schools even though my hometown had been on the top ten places to live in this country not too long ago (tiny little town). This reference is from; Rhetoric In Practice-subtitled Newcomer and Seward, By Henry Holt and Company, copyright 1905. Page 119 from the beginning to the end of the first paragraph:

There is another class of sentence in which two or more thoughts are embodied, yet in such a way that their intimate relation to one another is clearly brought out, and the effect is of a single thought. This relationship may be cause and effect, contrast, series, details of a single picture, etc. It may be expressed by connectives, or merely implied by position; so long as it *is **felt**, the sentence has unity. The final appeal is always to that underlying logical feeling called the "sentence sense," which, since it comes so often into play, cannot be cultivated too assiduously. END QUOTE.*

Cultivated too assiduously to my imagination is "to be figured out too often" when used in such a manner. The loss of their grip of the control of fear and guilt would diminish ever so quickly. Politicians use secondary sentences and agendas every day, and every election is the same. Most likely the position being opened has been to the candidate who spent the most money. This asexual political scheme is not blind to threats and I know I'm one of them uneducated threats. This can begin the cause and effect drastic action with force or violence, as the government has done so, so many times before, their cycle of Christian democracy and dictatorship of wars against another nation to have position over, and even their nations own people. People have been assassinated because of the causes they stood for are strong, and even if they had not been murdered like a tyrant, they had laws to have them stopped.

Now for instance there is this word we call science. This science has had one heck of a time imagining the things it currently has to claim proof of, yet it only started with a little imagination. Theories cannot be disproven until proof cannot be found and if you can find proof then it must be irrefutable and if even that is the case, not everyone in the science field agrees entirely on every subject or item of discussion. I'm now a scientist by that definition.

My exact reasoning for this style of book that asks not only my questions, but also those of the readers that will determine the use of their minds, even if they don't know fully what that means to their selves (no offense intended reader), so as to awaken the age of which man is to evolve and jump into the next evolutionary leap of technology and knowledge. They're the continued debaters of the days when the books of the Bible were dissected into which of them were to be in the original and which were to be left out. Science loves debates and loves to argue into stubborn corners of stagnation. That word stagnation is so far into the daily lives of every day human beings because of the sheer size of the debate being held and not forgetting the things withheld, that something must be done to eradicate this form of policy turned to a standard no longer reaching itself thus the lowering of the standard for more to reach it and so on and so on. A government is supposed to be in place for a few reasons. Those reasons are to let things flourish, protect their citizens from an outside force or threat, and to justly uphold the basic human rights every individual is born with. Things don't flourish anymore because too many laws hold people back, there is no fruit to bear witness to because all the fields are raped and not reaped though patiently we await a savior but when one comes like J.F.K., or Martin Luther King Jr., Nikola Tesla, J.J. Hill, etc. the salvation is selfishly taken out. Going back to my legitimate favorite stagnated person to mention is Nikola Tesla, the man who created/found the use of radio waves to send electricity through, wirelessly transmitting energy of not only information like the Bluetooth we have today and satellites in the atmosphere but also the electricity we currently use, though all of that was coming and not already in place, no one saw the larger picture greater than him in that era.. Electric charging a phone with no chord was a product available not so long ago, I wonder who had the patent rights, and as well as which person had the most invested. That technology had been discovered and utilized back in Tesla's day but the most defeating thing I look at is that Edison is credited for electric lighting being that of what it is today...no that's factually wrong because Nikola had made it from direct current(Edison's version of electricity) into alternating current(Tesla's upgrade). Edison is namely noted for creating the light bulb. Rumor has it that Tesla had died nearly penniless and had created well over five hundred patents alone. The radio in your car uses his works, and little is known of him because he was an earth mover and not an earth shaker. I must say well done Mr. Tesla for the creations you had vividly created for mankind's use though I cannot apologize for the misuse of the rights to your works because I don't support the uses of current political regime whatsoever in any realm, heaven or hell, planet or galaxy. So as to when this technology had become an issue for funds to implement free electricity to the entire world, the current financial banker he (Tesla) had was also that of Thomas Edison's, named J.P. Morgan, had only one question, "where do I make money for my return?" Rejection in the form of pride, the limelight was too evident to see then, though it's still not even discussed in as much as other less productive people like Edison who namely made the phonograph compared to Tesla's Radio. Edison's electrical sub stations to further the reach of his electrical output which wasn't even that bright to Tesla's alternating current using far less resources and manpower and time, including upkeep. If I were to have a choice as to whom I'd wish to meet in my afterlife as a newly returned energy to where I had come from, first Christ, second Tesla, and third I would leave up to the entirety of God's will. Just to revive the name Morgan, if only he had financed the inventions Tesla had created, his name would be as large as the creations themselves. That is no reverie, that's a rosy cheeked child playing at Christmas outside with a carrot and snow, new toys scattered around, with hot coco from Ma waiting inside. Life as we know it would not be the way it is whatsoever inasmuch as it is deceptive and disjointed, though a thought, and not lived through.

If you can't tell, I'm decidedly against the way Christ is portrayed, and politics are betrayal. If you believe His life to be ONLY that of a Martyr and becoming your salvation, I will consider you an idiot and not argue,...I don't want to argue with a fool. With his apparent wisdom and power, don't you think that the salvation part could have been put off within His will until his Eventual death? He is talked about like he has infinite power, though He had only one or two things on His plate? No, He had more to do if He knew He was going to be revered as such an accomplished man, miracles and all or no miracles whatsoever. He had walked on water through rough seas and when arriving at the boat of his disciples, he said be not afraid, and were immediately at their destination however explained. I dare not say he was a failure though I can't help but think that there's something missing to what I'm taught and it's not going to be told. This is the exact reason I think of the possibilities becoming self evident and beyond, only because I couldn't get a straight answer from those who had taught me their way, only to have an answer like "I'm not entirely sure", or an "I don't know", up to "you should ask so and so" because they wouldn't decisively be called out on the spot (for they did not know either), instead they made distraction like they had themselves been taught. Indeed if He through His Father had conquered death, then where is He, since death cannot kill him being that he rose from the grave alive and well? If He IS around, then why not be known and the ruler you're said to be? My imagination brings me back to déjà-vu. In this multiple choice question I will advance you the theory of relativity, and triumph over it with the simple symbol of infinity though I ask no question in its direct fashion. That symbol is our number eight which lay's sideways if you didn't already know. Most believe infinity to never end and the symbol to never break for its consistency. I disagree with both of those points to bring up this one. Ashes to ashes and dust to dust, the dark side of the moon has the rest of us. Well, not literally the dark side of the moon but if we are Ying, then our Yang is the other side of the figure eight and we can't see the other side of the river of life because the Mayan culture had done so, so thoroughly in mentality, which is where they possibly went, and why we cannot find much of a trace from their culture as I'm told there is not much, though I don't believe all the things I'm told. If however I am correct in the theory that the thing infinity is said to be can in fact be broken, then the leap of mankind can and should be done if allowed and not stopped from.

A theory of mine is that it is a natural ability to teach the next generation things sooner than before over and over throughout generations until the birth is united with speech naturally without aid of technology or drug, then leap onto crazy with me and we from there have in forms of civilization an explosion of technological advancement into the known universe and beyond the thing we call reality in our dimension or if you will, realm., and opposite that, the black hole within a galaxy within another black hole and so on(those civilizations who'd already failed.) Let's talk utilization of this thing a black hole and agree it is thin, (only to get this out into the open and off my mind) though massively large in circumference as we detect it from the known views we imagine. Now that were on a thin line, take for example quantum mathematics and bring the analogy of a piece of paper as the black hole. If you enter the face, you rip the paper or you get ripped by the hole, a fool's journey. If however you can digest the way to enter through the edge of the paper making it your doorway, then you are at the speed of light and don't need to think of the size you are. In another way to look into this is that of a coded message transmitted into and then out of the unknown hole, then it too is a winner of a theory. If you're past that speed of light theory unable to be surpassed, you may rip the space you're in for a glimpse at what it is on the other side of that speed until you can keep faster than that speed entirely long enough to collect data, slow down, and return to the time you came from if in fact that is the issue, time and the return being different. Though it is much too costly to consider such a trial and even if tried, still a foolish endeavor. The same is to be said of the super colliders which smash things together at great speeds, at an even greater cost to the economy through that thing of taxation. That is entirely enough of that nonsense for now. In another look as a black hole which is evidently a large possibility that science is unknown to claim as a theory they have, is that it itself is a civilization self enclosed and protective in every way imaginable to any outside force unequal to itself and unwilling to unfold the undeserved knowledge they have accumulated to those who have not the responsibility, nor the personal control to have such a control over their things appreciated as we would if we had done so ourselves.

If I were a complete stranger to you as I currently am, and we do actually meet someday, then we together will have a first impression on one another. The first impression is a most vital forward movement a conversation may use and not only the conversation, the relationship it can become. My first impression in a small in depth look at how I feel should be conducted by both peoples involved and is something like a game, and if you permit the improbability, a game of genius' quality. What is life with no fun like a game every now and then? I am considered a single person, as are you, so let's pretend that that added up still equals two, but because we are not using ourselves to meld into a greater one self, then other things must be used besides conventional mathematics. I, Michael Wolff believe one plus one has the sum of forty eight when not making a greater whole. Hibernation used mathematically though in abstract ways is of any value as long as that value is Thee value or of many values valued enough for acceleration off the map or "out of the box" so to speak. Place me in the nut house but here is why I feel this is important enough to share with you. The simple game rules of baseball and while at bat are the given of how to interact with another person when interacting for the first time or the seven hundredth. When at bat or in the dugout, you're technically in hibernation or in planning to make a move so to speak, and even speak correctly in patience and clarity. Take into account that the equations sum is two away from fifty, let's say on the way to one hundred, supposedly together, being that there is that gap to fifty, how should we proceed? If reason is kind enough from both parties then it is not exactly a race, rather an equality made a balanced race all together for the projection of one hand washing the other, helping each other reach a certain destination through a mutual understanding. One way of a balanced race is through questions as a way to teach through another their own fundamental appreciated values to be updated and not debased, if not already understood. Although Neuro-linguistic programming is an effort, it too is too far into controlling another's emotions to have a directly correct approach. The definition of a debasement- is an act of debasing; being debased. Debasement of coinage is brought out by increasing the alloy in coins. This same occurrence can in the honest perspective pre talking to someone, be observed from an outside appeal for instance to that of the opposite gender for your interest was somehow either peaked, or rather lowered and wondered why. If you ask someone to repeat the strike me once your fault, strike me twice, my fault appeal, theory is bust like it should be unless the appropriate amends are made. Tit for tat studies from a Nash Equilibrium and the prisoner's dilemma gamehad been found decent enough to understand. Meeting for example of two accomplished people who can consider themselves to be from the beginning more than the original fifty after the baseball rules are fully understood, used, and moved on with, that's called trust and were back to a pitch here. Tattoos are a symbol of accomplishment and some more recognized than others. Like a military ranking shown on a sleeve. The race of equality to one hundred percent is not complete until both people are at one hundred when an exit of one is to take their share of the work or profits provided, and not without prior notice to the other(s) because when done properly, no notice is therefore needed. A collective whole is like the sharing of breaking the bread, and giving to the needy for they are as needed as you, though you may have more. Now that a simple addition is covering mathematics in a new concept as mine has introduced, what is one plus two if baseball rules may not necessarily apply any longer as has been done so before? A different metaphor is indeed needed but if we are to retire using mathematics altogether for a little while then three together can accomplish more than that of two or one, the work load may become larger but the networked ability to get past the places a smaller group or individuals and easily found and usually used. From many pools of available knowledge, the possibilities are either deemed from the beginning the end result to be good or destructive. The metaphor for more than two is a gathering or a congregation. What is more than a gathering is a social group. More than that in terminology only, is that a formulating group can exist to find other ideas or people to add to their own original for independence or revolution. Revolution onto what this is in one example is education itself. Those who teach are underpaid, have over sized workloads if pay is considered, and still nothing to show for a squeak in the eyes of big government when the cutbacks hit the educated and education system as an entire living and non-technically breathing business if let free to expand its own horizons. An entire section of this new education ability and beginnings of how it could change the world is a complete section devoted to change and the business or businesses it can create from the students for not only their own profit, but the teachers pay and eventual self sustaining funding needed to keep the schools in progress and demand, all the way to the decreased taxation of the people within the nation for an easier payment to a debt owed. Let's move on for now.

Ending wars is a difficult talk and task for some, however if you respect this as a truth from somewhere you had not been, then it has possible applications in great many areas of life and ways life is proceeded through because you had not been taught any other way thus far.

A manmade material magnetically charged to that of let's say 100x's that of a normal magnet, can make movement enough for a load of cargo be it small or great, and be used in so many defeating ways, it's impossible to not use now. This new thing is the defeating thing of crude oil and the super high demand and stupid high prices it has gotten per barrel. Currently it takes two gallons to make one gallon of useable gasoline. That's way too unprofitable and unsustainable, as well as running out. A barbershop pole red white and blue turned sideways and then offset the magnets as they can be shaped into smaller bits to a larger piece of the whole, the stagger of this is also positively and negatively altered through the positioning and the powering effect if desired can be the moving part as in reference to the throttle, inside this tube of magnetism or outside it. Larger brick magnets to drive the smaller ones within the barber pole, I'm GIVING the WORLD a much needed cut, may it be called mercy or not, a whole different society is tomorrow brought to today and not a single death is needed for implementation. This is not anything that can be patented, it is free for anyone and everyone for experimentation and use, large and small to profit from, if not create for yourself alone and unhindered by laws restricting the use of it. The use of crude oil plummets for to the grease of the wheel today it not only squeaks, it screams and squeals at a roar begging to be proper like a nail head in the wood, safe and secure. You are very welcome. If continued as said in the use of this gift, the experimentation of, and use of cost effectiveness, congratulations on the first correct step in the right direction taken in decades. In any other argument that it must be made into a patent, I hereby patent my idea through my works and trademark the entire use of the propulsion it from the design proposed can be created from, thus receiving all royalties at discretion of myself and if desired, untaxable so to be distributed at the said price advantage of purchasing power rising to great new heights in this decade, and the next of what may be from said proposed forced patenting. If this by law is not upheld in court I may die fighting for it to be my own but first and foremost, free to all people. Basic principles of design are required in the temper to the design like 3-D printed plates holding but only one disc and alternating positive and negative until flush or spaced with the next accordingly for propulsion, continuing with the next disc, the additional discs are for more needed power distribution and thrust, speed, and breaking to slow down. The outside larger "bricks" are for their strength and needed balancing effect to make safe enough for massive speeds. Likely though I'm already in need of a high speed bearing not unlike those in massive generators to hold steady the balance the front of this that connects the mass in the middle to the probability of a custom transmission and so forth, and the use of an alternator running off the friction given, and quite possibly a newly designed cooling system for the speeds reachable and the slowing down of the said speed. If contacted to have this purchased from myself, be prepared to be laughed at and dismissed, or have an educated agreement made as to the production and price it is to be sold at, for I will have a great deal of my own aspects when the product becomes massively available, and the following ten years to twenty tops of how the price varies depending on demand. (The first one off the line to the end of said agreement.) as a roundabout uneducated/researched guess, the already stockpiled oil within this country alone would suffice to this ordeal of the design of cars for literally generations, leaving the processes of refinement to be improved upon and the prices be near the same, though the demand spikes only when the people buy this type of product for the dependability and cost effectiveness it provides to them as a consumer, lasting longer than traditional metal vehicles deteriorating over time and thus breaking down in need of costly repair. For those of whom are big into the oil industry, shifting gears with the amounts of currency you withhold should be but a smooth separation from one sidewalk pad to the next as it is called a step, even some would say a forward movement. And the long term process would have a slight increase in the decline of oil though not through this effort eliminating the need of the crude oil itself. The oncoming profits into this new era are for everyone in reality, not moguls who are entirely far too stupid rich that spending fifty thousand dollars a day on toys would in their lifetime not even dent their wallet. Out of context but not to be disregarding the jobs and displacement to those who have the training to recover this oil, there should be in place a way to have affordable relocation to those who are needing it through 3-d printed houses and intended communities of which are not only for those who are replaced by a newer version of transportation but also those who need housing right now, not company owned as before by industrialists, but by the people and their accumulative experiences and newly designed mortgage lower in price than that of a base model car which is today when in the twenty thousand range sticker price off the lot brand spanking new. In doing as much as that, the interest paid on this newer housing action through relocation is such a profit to the community it is as well as the nation because it is at such a low range, it is almost improbable to think of what it could be. The profits of any overage in a decided percentage, when it is made, can be put into effect this plan of renovation. Interestingly enough the current national infrastructure is crumbling and in dire need of the update proposed or another proposal of which would by my assumption be that of some politically driven profit sucking asshole or corporation. Utilizing the solar paneling and no longer the wind farms which I have a larger that the blue sky itself is, have a problem with. This problem creating the leverage to disturb nature by means of many levers and not just one needed as one before me had said, "give me a lever long enough and I shall move the earth. Not that this next item can be removed, I give dishonor to the Three Gorges Dam because it too disrupted the movement the earth had previously spun at. It when starting up had slowed the rotation of Earth by a measurable amount, and not that small either. Over time if you take the added penny a day then the next day two pennies and doubling again and again for thirty days, the amount is great. This slowing of the earth is not as dramatic yet, and given it won't multiply the same, the analogy is somewhat edgy as it should be. I mean nothing of disrespect to the country when I don't live there but if it affects us all, should it not be considered for all who will be affected? I don't like that it displaced many and even disrupted old lifestyles but the heritage handed down. Once again, it is not my place, and should not have another word to say about it though I make no promises. Why were solar panels not freely given to those communities when it would have been more cost effective, and the realization of the power they can hold not mentioning create the uses of individuality and more than that, the unknown?

Call me Crazy if You want

This may be off the wall left field crazy and bat-shit slippery, but what if the game heads up seven up were originally in the beginning only meant to be three? Not only the three fundamental original beginners, but also not as we are here entirely. From an aspect of groups of three within heaven and creation, the failed games are joined into other games called life and then the continuance is considered fundamentally unflawed. If for some reason it is or was started as only three to begin with then I'm assuming that one stays completely alive in body, one stays completely in the game realm and one stays as the mover who can go in-between the two realms but not make any real connections to the physical from where that singular person is. As one game stumbles, the connection to the rest of the games being the same as that of the first who had apparently miss-stepped, that which is taken over is the correction and a possible recognition to it. While taking turns at the individual seats, certain criteria are limited and others are permitted as the three legged stool of a group in this criteria of balance make their moves as they permit the one who is in body to make the progress further itself either from a backseat, and possibly to the front of the views of many people if even to fast forward the world into greater things like that of which binary code had been from analog waves. Radio waves are considered analog and even with that limited use, Nikola Tesla had found its key ability for use and was stagnated into a near penniless death with not a whole lot left to his name, and almost completely shut out of history as a name to be remembered. So this newly found code called binary is supposedly greater than that analog slow progress although very useful but not yet handy, but wait one second and you'll understand something that I've dwelled on for quite some time though stumbled upon by complete randomness. First to explain what I understand Bi-nary code is. A bunch of ones, and zeros, in a set pattern to represent from quantum mechanics to physics a thing or things that can be used as needed or necessary, filling a space or not in any size large or in the smallest of small. Adam and Eve had the forbidden fruit, and then were removed from the Garden of Eden where the original rivers of life began, being called a piece of the tree of life and cast into life having death for their punishment, losing the grace and light God had given them. We are taught through Christianity it is now a fire within us. If analog is the ways of old, like that of the garden and fruit, being prior the punishment, whatever that consistency may be referred to as through sciences and many possible definitions, then could it be that binary is that which creates our ultimate demise as a species from ourselves, or is it that justly thing where it still has the ability to be good overall. The fall of mankind is said to be imminent over a period of time thanks largely to religions forced onto a current non-believer. Entirely true, wars, genocide, disease, or natural death, it's all the same ending for everyone even if uninterrupted. Think, could binary openness be a universal interference of interface to those of whom elsewhere in this entire universe had gone before us? Was it their success or failure within the average thought of movement? How would we as a people find out this information? Universal laws are a probability though I won't decide to decide if that is or not, I know not if binary is being used in its probable correct involvement to humankind. I do however know that the Apple I-Pad had been created before the I-Phone thus the theory that our products not only get designed to fail or break after certain periods of time, but other products are withheld until they are made into slower progress for the consumer and greater profit to the company producing what it is we consume, always the consumer following and not integrated with. Possibly it is another test of God to his creation of man to see if they are in fact ready for the universe to open to the welcomed and not to the damned. I won't question rapture, I only question those who attempt to guide complete national acceptance if not worldwide forced acceptance of a criteria of down your throat control and nothing to do about such a thing. From those in the three person game of heads up and having movements of their own in any of the three areas to the tree of life, even if an aspect of abstract explanation, the one in the current body can in fact die ultimately overall including the game and its design, so then what of others who are in other groups who had not succeeded? Becoming greater in numbers, within body and or lines forming behind those still playing the game awaiting life if not the last one still yet to fall into failure, the game of what I've been told God plays with mankind is called Lela and He the Almighty is the only player, and the only one who knows how to play. So the theory of the three is a possible failure in its theory but what is imagination if not a little fun with the gift of choice not taken away? Is there anyone taking sides?

Time and experience!? Can one be without the other or does the answer befuddle the babble from what was Babylon as the story is told trying to reach the heavens with their creation from their own hands? What is nature versus nurture to me?

As a rule, we disbelieve all facts and theories for which we have no use-William James, known as "The Pragmatist." Is reason more reliable than experience? Pragmatism is defined as-an empirically based philosophy that defines knowledge and truth in terms of practical consequences.

Chapter Two as a tortured mind continues

As from the same Holt and Co. reference as from before, though I don't believe they'd actually have the rights to what this explanation is, for it's just that, an explanation. However, I don't want to get interrupted later by chance for inadequate misrepresentations and legalities. Ch. 2 is Description, and in this reference; 'The primary purpose of description is to show how objects and scenes look to us. This purpose can be accomplished best by means of pictures, and wherever it is the sole purpose, a description is simply a convenient but inadequate substitute for a picture.' Description is as they continue to write, an art in and of itself independent, and arguably the most respected form of. Now to what I've been understanding for a long time is that an artist, a true artist, is someone who portrays their current society, situation, or thoughts of the proceeding current events as accurately as possible, and in what may or may not be a clear representation to their method and definition of why it was what they made, made.

Readjusting the systems management, from a leader's point of view, if I were a leader, today though far from the acceptable normalcy in a psychological aspect, I am still capable of functioning in everyday society with matters large and small alike, the same as almost all those who are within that same standard of what had been before them. It only matters to what extent of the things I take responsibility upon that determines me productive or a drain on the rest of those around me, and the society I live in. Another standard is that of the test of a person's intelligence through questions deemed more appropriate than those less likely of things like imagination to which is mostly a perspective and decidedly arguable to more than most. Nirvana itself can be added to because if it is nothing, and nothing wanted or given, the understanding is but one step closer to finding out what feeling that itself is to appreciate it as an honest euphoria, like it should be. Just to understand something is not to desire it, only to uphold it within knowledge and the understanding of what it means with the responsibility that it might need to uphold such a thing, raising the standard of first the self, then onto the ones we care for and so on as a position of ripples in a pond reaching further than that of oneself. The addition to what Nirvana is, is not only the representation of how you proceed from it, but also the leading up to the scale of thought that determines the euphoria and its self administrative feeling of which you leave your mark on for others to feel when they themselves get there if at all and the length of time this thing will linger within you and or around you. Meditation is the way it once was told that was the only way to achieve such a thing, however through only deep long concentrated thought had received the acceptance of this gift, or may I say interruption unto my thought and life, from that point and on, have looked back many times into memory and found numerous things as aids to my thoughts and daily life more than anything else I've been taught to depend upon. The way I found this was presumably not taught to me directly, however it has me wonder of the other forces that may have guided me to that point. Proper guidance within my control had most likely been the type of music listened to even though much may have been a distracting thought, either way had been needed. The after effect to that degree of future undisrupted thought and concentration is quite possibly the thing that has me carry on with this diligence. I know I'm propping myself into this clean lifestyle though it as I write it is nothing close to my desired living. A One bedroom rented apartment and most days nothing of currency in my pocket. The day I get a job is the day my provided insurance is cut off even if in the middle of a process like a dental procedure needing a partial denture. This is an extremely hard thing to crawl out from when not given the proper chance to excel. If though I keep myself pre-occupied to the things I could call a passion or even temporary relief not drugs or alcohol but as for myself reading things making myself more informed, thus more powerful, then I have still a more clear mental picture to what end goals I move towards in forward movement than that of what may be the fog of war, the clouded mind. In life that thing in which thought is that thing in the way designed to be but one distraction making my mind wander aimlessly after varying uses or addictions taking over what might have been my moment to find importance to something no matter where or what that importance may be, though in the fog, hidden and hard to concentrate on tracking the tracks previously tread, I remember the fog I had encountered and now that I have been as medical analysis would say less in the fog and more able to aim correctly at a goal of certain depth, move with more ease and ever growing clarity and less disjointed thoughts and movement.

Chapter 3

If you have the desire to understand that Jesus was a man and *still* the miracle maker he was, then I have a portrayal of mine you may like. To have those who He sent out, going to the first homes they were invited, told not to wander from house to house and that someone would invite them in to stay, and stay they should. Now permit me a little leeway to the way you have thought it had happened or taught to you by an elder as you were uninformed at the time of hearing it for the first time. This is my first time in explaining it to anyone, therefore you may become one the wiser as the first of which can claim were "in the loop". First, Jesus had his hand raised unto his disciples forehead and merely touched it having the individual not only learn how to do this, but be able to do this at the place he was welcomed. Now to tell you a simple touch had to be learned is out right bland and meaningless if however you do not consider in what that touch had been, there had been many things unseen by any looking to this from even a rooms distance away. From the instant touch had began to the instant the touch had ended, in between for the recipient had lived a thousand lifetimes lived and none dissatisfying or unjustly, the question had not even the need to be uttered because it was true that it had just happened, and as well true that the lives they lived were of not only importance but as well, real as they were. As the touch had ended from that simple moment been learned and given as a gift to those he had to send out for safety, those people who had invited them into their household had been given this same treatment as a testament to the power of God. All along the lifetimes lived as near a dear brother the gifts to the heads of the household were only to tell them the ties that the days they were living in a sense from Christ would never end. Not in fear these people, but in joy because if you can imagine more than a handful people within the world capable to destroy the entirety of it more than eight times over, having the power to do such a thing, yet only to return to their livelihoods again and again until their realization had been uncovered, because with it being finished or destroyed, then only made back into the days that these gifts had been spread out at a seemingly random locations and in separate ways we now live and can revere to Deja-vu through Christ and the many connections he has to insure the survival of His Fathers creation and all the things within, would also defeat the devil as a slow and painful nightmare unending and relentless. So relentless in fact that it had the deafening tone of silence of defeat to the liar and his deceits. Although disabled, his first deed of unjust disobedience was not so easily correctable; there were still many ways in which it can flourish from its correction through Grace if even made into revelation- the act of revealing or a way to reveal.

So, as to tell this story one way or the other being that God has seen this come about and go noticed or not, taught or within the forgot, you know there is another guiding force of creation and it can be controlled. That guiding force in bad or good, controlled for the desire in effect to oneself or another is an imagination, or a subconscious drive or feeling. Not to overlook the ability to have an imagination overrule another or misguide another, it can beget itself one way or the other, pleasantly or not, as in the same way the story is told or retold differently. Another reason I feel it necessary to repeat that $1+1=48$.

Dominated by religious stories and the belief they portray, check and it works out with these variables and more. Why if any would there from this, be any static or retaliation either upon myself, the people, the rhetorical question of what is to come from leadership, and the future generations leaders if not the perception of change if it had happened, or not at all? This is not much more than what has not yet been termed as I believe it could be. Quantum thought-that of which has no limit upon the imagination primarily used for rhetorical theory to oneself. In quantum thought, one questions without looking to what has already been itemized or written by another theorist considered the current standard,(definition by definition is still only definition created by another and slightly accepted as relative over time) taking into your own control for instance the question of why we are told stories of people living into the thousands of years and living that way peaceably as the side note. Is that only because when time had been first measured, a constant drip of water gave the illogical inventor a pathological thought that he must make it a reality to have time be of a criteria, thus a standard? When the two figures of thought connect, will it reveal the answer or yet another question? Bring in another suspected thought relevant to the situation or not relevant whatsoever and there you have a path to remember by way that has you yourself remember the oddity yet relation to what had been thought in precise form. The difference of forgetting a simple thing versus the recalling to what is needed at the time it is needed is a line of integration, not only concentration. Little practice is needed for this to be done correctly for most who start can find this form fun and intelligently forthcoming.

Mentioning the sum of forty eight and where it had come to be from, the reasons having thus far been, innovation, justification, logic, andthe search for more. The 'search for more' is very close to 'what is more' such as philosophy. What it could be though, is together the same thing with different outcomes at times. A personality of a thought can be transformed, or created altogether for later use this way in real time, the universal way where time is almost nonexistent. Pose the right question and bring it to a real world mathematical formulation, commutative and contemplative thus the transition of the formulation to the real world formulation. I think Mr. Einstein would be proud this is still in use whatever the way, creative either way, different and not the indifference.

New Chapter

I now bring to your attention that of Depleted Uranium. Every year reportedly 50,000 tons of depleted uranium joins the already substantial stockpiles in the USA, Europe, and Russia. The accumulative world stockpile is 1.5 million tons. Some of this depletion is used to dilute high-enriched uranium released from weapons programs. This u238 is not unlike the same toxicity as of what lead has. Reports say its radiological effect is safe though relatively dangerous if inhaled. Now if you must, go back to what pages had presented the magnetic non polluting environmentally safe power of thought and innovation in the simple analogy of a barbers red white and blue barber's pole signifying that a haircut can be gotten there. If the eventual goal is the solution that gets emissions to zero, scientists should fight for these innovative things with or without the funding...go to the private funding and forget the regulative government choking the life out of life itself which isn't dreaming anything but the nightmare we all forget to acknowledge. As a self leader, I bother nobody but those who attempt to dominate a society they disassociate from. They who do that don't feel the daily struggle because they feel that they are in a balanced state. I say those people are the ones off their medication and should be forced to prove their need to not pay the same taxes as everyone else. A fortune five hundred company has not had to pay one dime in taxes thus far and this hurts the entire whole as the whole mentioned is the existence this planet has on and within it. So as to why this gift should be free, is the simple retribution that the past cannot be replaced, only added justification to from a corrective standpoint, therefore the gift is free to all if not argued against the ways in which have been stated as a reliable cheap automobile with the price affordable to what today's poverty could afford in the year

2014 A.D. I see it as a fair trade to the problems with which are formed form inequalities in society. Not limiting this innovative creation ability to only automobiles, how's the industrial industry going to evolve when the power grid supposedly needs not only updating, but more power? With this in effect, other ideas like this can and could continue to reshape the face of life and how life's face is endured to an individual aspect of self introspection. The word relative is of many uses. I have a relative, which is of a relative thing or instance, and can be relating to another subject entirely. I heard that there are over 500 words in the English language that have over ten thousand different definitions thus uses. That is only the English language that has twenty six letters in its conglomerate of possibilities. If that same simplicity of uses are within society and actually beneficial to the whole and not an elite few in fear of being outnumbered, then possibly the chance to move forward and not repeat in repentance by God's command is within the grasp of mans own hands.

New Chapter

Bring new and exciting elements of creative thinking to the old mundane routine rut because the stagnant water untreated and unfiltered is as deadly as guilt. New thinking can be intimidated and bent to the will of the user. If and when many winds of the mind won't stop, one of two possibilities exist with four different scenarios from two sides of a/the scale. The same thing over and over is considered insanity and nothing as far as I know within society besides the computer as a major new technology has been introduced recently, but computers have such great advancement speeds that it must be introduced to other such advancements so to have proper balance additionally created upon itself if not as another complete whole in and of itself. Just as a projection is, it is not an invitation to be interrupted by another projection by or from another person or thought of individual perspective. This is a stalemate today to the future of possibilities and the control being designed against the average individual if in fact some action is that of interruption, and who really wants to lose a loved one to a fight or war of which is unneeded entirely by the rights all humans are born with. Where within what destruction is the correction? A correction of proper content is the redirection or redesign with acceptance, not through force. To replenish an empty stomach you first must find what it is that it will be filled with, and then choose to chew or not. To have an echo from there, when a body is not replenished sorrow follows soon, and after a long time the wane of death lingers ever closer. These things within the pages of this book are the David and Goliath of our time. An artist's sole purpose as an artist is supposed to be a representation of the time they currently live in, though blind or misleading to the blind, today is not divinity lived as through the pages the magazines and movies portray. Entertaining as they are, they are only a small update to what history we as a civilization have had to endure. Fighting, fighting control, slavery, and the hero aspect and that of the persons listening or somehow learning of this hero try to live that story out as it were them is an inviting yet distracting thought. How was it found to be another use of control over another person? One can only guess at what that was. I myself statistically think like a lot of other recorded schizophrenics and believe in angels, demons, and narcissism, hear the imagined voices of what I don't always understand, or necessarily consciously hear as well as have seen things not within reality relative to things of which I myself can even grasp the understanding of for the desire to do so is far from my goal.

Snowfall itself, mountaintop or main level at sea, any season within what it provides as a majesty over the preceding one, the next and the other, is though an imbalance but also a neutrality to have itself either a friend or foe, least in itself as well as from afar another season to it in response. Man and history in a cycle or repetition is not unlike that of nature and its most valued repetition though going unnoticed. Preservation or perseverance wins if it fights itself though it is itself a balance. Introduce the other factors that may be useful to either one, the balance no matter the speed of one or the other becomes faster than the other. A game has said "I'm so bad ass, I whoop my own ass twice a day, and I win" and this game says it in such a way it actually makes sense if even left or right brained, fought the other and even itself. Bits and pieces may fall if chopped and not sliced for memory to recall.

To massively speak at birth it must be handed down with great care. Immortality can come to a shared understanding as the word is, but to be practiced, perceived, and put forth, patience within all sides is first understood and planned out. The debt felt within oneself being the patience towards debt and or doubt and what could become tomorrow, the country, and or at least the person(s)/people within it. The people within the place are a form of wealth because who would do the thing required for the thing to be done if not for the people to do it. He who does the thing has the power, he who does not the thing have not the power. The debt America is in is consuming our tomorrow and war is an ever growing fear on six different levels. Two countries want to flex their powers and one we have an incredible amount of debt to, within our own land we have the well designed unending wars of drugs that make them more desired not only by the user but the usurping government itself as a form of for the greater good we will continue taxation, and the so called war on terrorism in which that too will never end if nothing in the greatest form happens to intervene into it. That to my counting ability is six if not more when looking at the daily lives of the problems that from the top rung of society drain the people who lay underneath or attempt to crawl out from under. I have careful consideration to what nonexistent leverage there is today, but this season in a form of debt, is unlike that of the far off one we know nothing about. The question suffices as many as many may be while the answer drives the pursuant one to make greatness greater. If when on a journey you could use a post to lean on for rest, momentarily pause, inhale, hold it, count to three, exhale and re-analyze or re-adjust if necessary to begin to better society as it too is now your journey as it has been mine, then we can work together in my opinion. Sure a wise man can learn from a stupid question, but a stupid person won't always learn from a wise answer. As well as that can be, the ability to admit when one is wrong, and even greater than that, change that which is wrong to what is right, is of the hardest things to undertake from an outside perspective to that which is right and not unjust and inhumane if you are part of the problem.

New chapter

On an entirely separate point of view, if I am actually Jesus Christ, an Angel, or a normal man, and even all four placed within one body sharing the soul of grace then as I live today, is it One; as in the touch to those of whom I had sent out, living through the hell created and saving souls as I live through hell as well as judge the ones to be on the next life in time of the next life I live in hell or is it, Two; isolated to a degree of comfort based on the purpose of planetary success not being that of my own? Or Three; which is sometimes the scariest to look into, all three of the mentioned divine powers at once, all within the same body, sitting along the riverbanks of life waiting for that something big to happen for any of the three of us to have the complete and unedited version of self as the other two within me leave to another extent as the extension to be called upon when needed after that something had happened. Possibly just a normal driven man though, I do not consider myself as another would see me. Arch angel Michael used as a reference to battle the people instead of Lucifer himself as he has already been slain, Jesus the Son of God awaiting the end for judgment to be what it is intended to be, and as well as finished with His work, or The Almighty for the reigns of patience to the other two and if necessary when over, to build his kingdom as desired? Now to introduce the idea that I'm no Napoleon but if I were to create a kingdom from the ashes of another which had previously fallen, then would there be any psychological effect to thyself for its failure due at a certainty as the last had fallen or made their way and intentionally creating their own kingdom in no need of direction or war in massive planetary destructive ways as still in our own? I'm placing myself in others' shoes and trying to find a pathway of thought that is more ordinary and quite possibly more acute with acceptance the subject of clarity the matters brought up by my curiosity and satisfaction which are thirsty, may be entirely out of place, though initially within reason and not without faith.

    To start with what this is about, for some it may be a leap of a new consequence or a leap of faith or a leap of desired acceptance that they're not alone in the same pattern of thought. Begin with the definition to Pragmatism- An empirically based philosophy that defines knowledge and truth in terms of practical consequences. William James was known as "The Pragmatist" for a notable quote-"As a rule, we disbelieve all facts and theories for which we have no use." The day is August ninth two thousand and fourteen in the year of our Lord. Is reason more reliable than experience? From reason to reason yes and no to a basic extent for the amount of drift reason or experience has merit with, for, and possibly against the argument at hand. The forthcoming writings are, as of what I've prayed for and now can call that of my own. Though the facts are real and I have not prayed for these altogether, I must place them within the experience a reader may relate to so the writ has merit. I pray to the Almighty in the highest that this book is from His knowledge and ultimately FOR Him and His works as He commands. No matter the strength of any man, all the way to every man, the laws men have made insomuch a way that by their law has made money their "God" through the root works of two men named John Locke and Adam Smith?, their strength of themselves is nothing in comparison to the one true God. Understand, one man made a law not knowing the second but expected within time it were to be another to work from his own works, that would inevitably cause the collapse of the modern age at the time it would be its peak of destruction, and was in fact done as such. The backed by gold standard has no meaning to any single dollar in the entire history of mankind if you make it itself the God man is to worship. Peace be with you my friend has its own patience to sink into the beheld, but the immediate gratification has no lasting effect any longer with many of the poorest and the richest choking them from that which is given by the breath of life itself, and a right to every human. Over time the use of backed by trust has made time unavailable to save our souls from our own works. The society in which I live in has Greek roots of philosophy within its core and into the outer reaches of any of its known or unknown domain. Which includes the top tier of society all the way down to the beggar's in the streets in need of more than these societies "Plan" is to "help" them. The rationality that "might is right" no longer applies to acceptable living standards and I will explain how wars in all their power are mostly unneeded for their majority of uses like the oil industry or corporations that pay not one single penny in taxes, and who also have more voting power than any and all of the countries voting citizens added together. This standard of living ((in unjust fears of a dictatorship no matter the name or the oncoming one under a cloak and dagger, ruling relentlessly with those who want the right thing to not to be, and the wrong thing to be, can be changed for what it should have been even since before Jesus' day and age of the miracles he had made.) However I ask what you consider a standard? Something that has reputable ripples throughout other acceptability's from generation to generation, or is a standard the accumulated amount of what's been as an average from before? Both of those acceptability's are not acceptable. Though both of the latter are what is used today for rulers, neither is in the right though they are one hand and the other of a serious dilemma that has attempted to maim and kill whenever it so chooses, or consequently begins to feel threatened. One hand and the other can do so much for one another but at times do just as little for the opposite effect. Another notable quote is that of being "Man will make money any way he can. Dishonestly if he can and honestly if he must." For the depth that that has to be understood, it is not only those who lead but as well as the rest of those to whom they lead. Mainly the "might is right" application is people being lead away from the good, and the right of life to be their own self power, or self leader and or for their own plans, unhindered by this stagnation chain at the neck making most people live from payday to payday or even in poverty and beyond. In that respect, the inclination of decline that has been honor, respect or daily living, purchasing power, regulations over what is or isn't more or less acceptable for daily living that leaves humanity to deal with the consequences of its own failures not only from the leaders of society, but those who don't even bother to care, is in such a decline that not just anything can be done to re-route the course that will be our future. Our created history in truth from today and on are told stories and within textbooks to which our future generations can look back on with virtue for their days if the needs of that what change calls a need is done. It will have to be such an incredible movement, it may be believed it took more than a normal man to do so. It may in all factuality, take every single man and woman to succeed. Faith in the past when it's been told to us as a given guilt stricken life, has been lead as powerfully as possible against the favor of the people. The next section is more to do with my creative mind using the availability of telling a sort of story that may possibly be used as a template to oneself, or a base of what was used to have a relation to chosen personal company, and how it had been held as it could be considered a living thing and moveable thing. In a similar way as Jesus told the story of life himself, I have sought the same.

My name is Michael David Wolff. I was born as I believe it to be June twenty sixth, nineteen eighty six. If it was to be expressed in numbers it becomes 6/26/1986. Not to disarray or make fear the means to which this is hopefully used, I have to explain as best as I can for what that previous exclamation is to me. Notably the number six is commonly known as a cursed number. However if that may ACTUALLY be within humanity's definition or understanding is beyond my grasp for the simple fact that not only do the spoils of a war go to the victor but also, the pages of history and how it is to be known, then the number six if in fact the victor be it God had triumphed over the devil, I don't know why society has used this number as a reference to the defeated. Jesus Christ of Nazareth was a great man in his day is how people know him to be. Its whispered that because Jesus' Father, our Lord Almighty Alpha and Omega raised his son from the dead and back into life itself, that Jesus' life still to this day is more than just a spirit among us and has a body undying. While in a quiet rebellion against the continuous degrading leadership of the so called "Elites" who rule even to this day through the force or threat of force mainly for their need of a tax, to live an easy unjust life, what they have deemed a value or a disvalue, a law, tax, or needed war for more power or control over another nation, and at times against itself, just for more darkness to overwhelm the congregation that their nation is already in fear of, as any gathering of any peoples is considered a congregate, there must be a point where this Jesus character knows eventually when and how to disassemble the portrayal of not only himself but as well as the future possibility of some other perceived God so man may make great advances when needed, and not be held back. What I believe to be true within today and completely overlooked by the use of greed and money is that a nation's wealth is not only in the currency it uses but in its people and its people's minds that can be greater than what they would be in chains as metaphorically like today. Even from their own hands can something great be crafted at even the youngest of ages. Though many jobs have gone overseas for cheaper wages to third world countries and other powers, the crude oil consumption is two or three gallons of crude oil just to create one new useable gallon. Now for the first name I've been given being that of Michael, an Archangel of seven who is considered the Patron Saint of all Soldiers, I believe it also to be blessed because of the Arch Angel factor and abilities. So to add that name to the month I've been born I have an undying faith that it overrules that ugly number six. Not only for that month having been beaten but as well as the devil himself who roams the earth to and fro here and there waiting for his next meal of sin from the living. For the middle name I had been given, that as of David also blessed, who as I reference the son of King Solomon that had written the book of Psalms through song and poem, logic and understanding, also defeats the number of the day I was born, on the twenty sixth. It overlaps itself for a reason of code and its repetition of the contents writ only a little different in the next mentioning of the same which had been before. The last name I have been bestowed is that of an incorrect spelling of an animal that is rumored to be a cursed animal as in the other way a blessed animal to other cultures it's also considered the king of the woods in other cultures. That cursed, if you will call it that, animal is a wolf and how my name is spelled as Wolff is not another analogy to something beautiful but to the extent of what may be considered a box of darkness at the moment of being too late to fix what is here to be fixed. As a dangerous animal alone and even more dangerous in a pack, this vicious animal is in retaliation to the possibility of the failure of the planet and its demise. Not only do some cultures envelope the wolf as a creature, but one of mystery. This box that has been given as a gift is as well a gift of great understanding. What is pain with no pleasure? The suffering has been too long and too vast to not fight for what's great and available today. Though my thoughts are to which I am certain are my own within the family I had been raised, the only great story (As I like stories) so far that I can recount through memory of having thought, is that of a single stalk of wheat being approached by a wolf for the first time many ages ago. This wheat stalk asking for God to protect it through what was to its knowledge going to be its last moments, heard nothing from the Lord. The fear may have grown as the animal neared, but as it brushed past and beyond then looking back to say to the wheat, "you're more needed than I my friend", and continuing on its own journey is a powerful mystery to this day in my mind. For the wheat later to speak with God who had watched quietly, asking if ever there were to be a day in which it would be in the form of a human, but not just any human, God's reply had been "There becomes a day when that is needed." Over time the wheat and God had debated here and there and when the wheat was mixed with the wolf and then also the body of a human, the unusualness was far greater than what had been expected by even the Archangels. This gives reason for the extra emphasis within my given names spelling. Possibly I enjoy my creative mind like one would their own voice, but then again what is the use of a good jest if one cannot take it to heart and understanding from themselves, to themselves, and then grow from there. Life itself it is and always has been a miracle, even though not all life is human or of blood.

A three legged stool with each leg having three more legs as in such a way that it most of all resembles the tree of life itself (God, who created the story of life to everything in it), that wheat, wolf, and human body form for an angel is only one of those three legs and to be certain of it as only the first leg is unknown, and unproven to this day. Still not yet branches but more like the legs of an unusually large stool, the other things my powerful mind proves reason to, is an ever growing understanding to what I'm made of besides the ash I someday may return to. An Archangel is yet another, whilst I resemble the body type and what is perceived as his facial structure of Jesus Christ himself(though throughout generations and genetics from body transfer a shorter body and less lengthy face) and for the third considered, a possible anomaly, a natural human male made by God to hold all three as one within the three legs' beginnings, the three legged stool supporting legs while the supposed cursed wolf is the last resort of myself and the cry for help from humanity gone too far (the box of darkness as the gift, though I know not what the encompassing of its entirety is, for if that were true then Jesus would be the seat for the legs). Why do I think this? It may be that I am wrong but other possibilities do however exist to why I am named this from such a birth-date. For I do not remember how, where, of from whom I heard this next part, I still believe it to be an account to consider. Archangel Michael stood before heaven and asked all the other angels deemed worthy of such a fight to restore humanity to its correct path and none had replied as they would join him, came on his own as the only Angel in heaven to have enough love and courage for God and His creation to enter into such a dark and destructive place and attempt restoration. The kingdom of heaven is within us all, as is the story of life and the ways it can be beheld, so why cannot it be taught or learned to be more like Him as it is already taught as the shadow factor and the anomaly? One point so far for the past leaders and up to the current for having that a lesson which remained but as well that has stayed one as well. Michael on his apparent own from heaven armed with God and the other parts of hidden knowledge is here to fight for the people crying out for the change to come and not to be left "as is" "life is hell". What of failure is another question pestering myself as of late until successfully found, is the pretense that victory hides itself within, and away from those who chose blindness!?

Restoration is the wealth of knowledge that so many pools within life have, and have not yet been gathered or given to the people as needed from, or attributed to one another for greater human life potential and lifestyles of advancement to the mass populations in most congregations willing to do as Gods laws of life, and how it is to be ruled (even over the living animals and plants that were to be taken care of by mankind) is only one of my greater questions and quests I've sought out for organizing the world to return to Gods worthy love, or entirely shut out from use for previous mentioning of misrepresentation. I mention all the above for a simple reason that may take an individual a little longer to realize than others (minus the fact that it's usually the length of an arm or more as writ). Jesus Christ was a teacher of the story of life, especially to his disciples. It has been said that when Jesus took twelve ordinary men and taught them for only a basic two hours, afterwards they were able to speak directly through Him as though it was from Him because the individual He was, was able to make those ordinary men into great leaders in a short amount of time, speak and teach as he had. The apostles he sent out had His thumb on them for only a second of time but had lived over a thousand lifetimes by His side learning what to do and where it could be that it would be needed as to hide truth from the evil man cannot release, was also taught how to tell the same to the house that had invited them in and further the safe zones for which we would need because time to heaven from earth is slower, as is earth to hell…the farther you descend, the faster life becomes, and Jesus though we believe it to be only three days in hell for the toll of our sins, it won't be a specific number explainable in words. The men who learned the second touch not directly from Jesus' thumb had to know what of Arch Angel Michael, and the astonishments he would have to make possible in his own right by God's will. Taking this truth away of possibility if done, is over but I ask which are you in of many and where will you go if your opinion is wrong? I believe that the message He was trying to spread through his miracles, teachings since he was a young boy, and the travels he had made, was overall turned around by the "victors" who made him become THEE sacrifice of mankind making the worshipers the guilty ones and cursed through fear and passing it around like frightened mice. These safe points are the relics spoken of but not in reach of a physical aspect, only in heart though of who's will you descend to or from? The time of this day may be of many like it, all through another man's touch from a teacher using collections of these heartfelt consciousness'. Our time is not our own is said to us once in a while but if you stop to think about that singular statement then this sort of deep thought is necessary to the belief that more exists than a little story of a lamb becoming the creators sacrifice to mankind as the only reason he was here. Education suffers for not teaching to think…am I wrong to tell this? Why? Guilt is a very fashionable power and useful for control.

Power and the control of so many people is an extreme addiction which is quickly diminished for the next to be on the top of whichever hill it is that rules the most land or wealth or advancements, though this lands work has been made to be put together outside our own country as an outsource, then shipped back. As a reference to the addiction that that has been, in today's society an addiction like a drug is that of what we today call "The Limelight". The limelight is also known as a person's fifteen minutes of fame, even if it lasts a lifetime or most of it and all the way down to the fifteen minutes. Guilt though is a valued attribute to the bible verses throughout. To be thankful for and learn from such a valuable thing is a grace God has left earth with to learn from because guilt IS a mistake. In the book of Job he is given a test and had succeeded. Test number two was even a success to his Holiness because of the proof a blameless man had such faith in his God. Even though many places and people are tested, today is not a day to roll the dice and expect the five or the eight to get you out on the play date.

Here is a section that always has God the victor and the rest of humanity as a whole the suffering yet still rejoiced because of the greatness God is. Why you may ask that the people had rejoiced for humility and suffering is justice to the way I think at times and will try to explain. I know I'm long winded on paper. I will offer no apology to that. Whether or not it had been Jesus' disciples or His Apostles I do not know. What I know is that I have the un-doubtable question as to the direction that the book of the Bible had been written apparently sixty years after his death. I fully understand that it was from the word of God his-self, yet I doubt his word was for all but a few select people to be monumental in the pages of history, and also those of who are rich beyond belief. It is possibly that because the time the book had been written (After His Death) that in fact the message He was trying to teach and not just spread was turned against the very people His attempt was for, by martyrdom and the sacrifice guilt grips all believers with at some point. As history is a cycle and at times repeats itself, we as a people know that the best of what we have had as helping attributes to the larger scheme of things is stone cold dead before the work is finished. Our greatest need is being TAKEN right out of the pages that are to be written as the history we should really have. Now for the re-write from someone who knows his death is imminent for his works to the people.

I don't disagree that Jesus had sent out a number of people and having said to stay at the first house that welcomes you in as their guest, and do not go from house to house after that. That in it may even be a fallacy, however now is where I differ in the elements that had occurred within those homes.

Now directly speaking to the reader, if Jesus could heal with touch, and somewhere within his teachings taught those whom He had sent out, one thing to explain to the head of the household, then as a guest that was to be their contribution for the welcoming in, was definitely this in my mind. Every single person sent out were strangers to those who welcomed them in but only for a little while as it is told. A single touch to the forehead would send the receiver more than a thousand lifetimes into the future yet living every one of them unafraid, and with he who had touched him by his side for the singular purpose as to keep the reminder that time from that single touch had not yet been released and for teaching the way of fulfilled livelihoods, and much more. For example touch your arm for a second and then release. That is the miracle that is within my mind and why faith is in some, SO strong, that it is undeniable that existence is more than just existence. What I refer to is from the way it's already been said as "Existence simply exists". This belief is of the strongest in my mind, and is open for debate in a nationally recognized live setting with enough time for people to "Make time for it". It matters not who it had been that was sent out to knock on doors until welcomed in, or however many. It only matters that within justice and God's plan is working no matter the religion, God is God no other way, or from any other alter. As a matter of what you yourself are simply raised within, (which religion or belief) is most often that which you may continue to send out into the generations to come. As for the way the bible itself makes itself the most prevalent around the world, it's little known that it's shoved into the faces of whatever land it reaches. Sugar is a powerful attribute to how that is these days, and it is unknown for how long it's been like that. That shove Christianity is, is acceptance of it or death by sword. Now can you say a monopoly of that so called "Holy War?"! Is the definite core reason to other monopolies within corporations? At first it was the grain, then the newer and newer technology, be it an axe handle that swings better or the better and better monopoly on what was a tax, and into greater and greater wars. WW1 and WW2 (World War one and two) were the latest and greatest because the powers that be were "United" for a, or the, just simple cause of a "Tyrant". Need anything be said for America's first Label as "Owners" of a land they did not possess? I agree they had done an injustice to Natives as well as others. Nothing has been done for that to this day besides some political spit in their face. LAND TREATIES? What has been done is definitely not enough. Our complete system is a tyrant to not only itself but its inhabitants. If ever there is to be this desired debate as requested by its openness, then I am fully ready even this day that I'm typing out what is possibly printed. I'm an American and proud to be one. I'm however not that same American who is proud of our politics, nor proud of much of the past we claim as justice.

There is the debate that is obvious yet impossible to those in power because power is their addiction as well as their "brotherhood", lifestyles, and anything they feel is better than the rest of the nations average and all the way down to the poor, and of which is not anything to be proud of no matter who you are. In those peoples defense though I see that some of these people have chosen to live away from the system of society as far as possible yet able to survive off the more literal scraps of society by choice, not always alcoholism. Two things have been said there in the thoughts infancy but more are definitely in line for their breath of life and exposure. So go ahead and party hard, but remember that all it will take is a touch of the past to be re-set as the stagnant slow life and no electricity. The obvious debate is this: knowing that within the Orient there is an honor to their fighting of which is unlike any other on the planet, that not too many will go against their discipline of life, but there may be this option of declaration that IS worth a chance. That chance is to have everything from the bottom of society to the very top echelon be over thrown from power and handed to the peoples that this land was unfairly taken from. Call me insane but no matter what our banking system, let alone any other system currently used is built to fail while NOT in the average citizens favor. Free bank loan interest, giving a slave rights, and electricity that we pay for yet could be free another way (ahem, Tesla) and other things like proving the American government they were wrong to call Hill's work a Folly. Our little nightstand and bible stand against nothing but the cold. Not to mention any particular country before another, the orient is nothing to mess with, period. The way we as current citizens are fouled from America as not the founding fathers but the law it had become over time. We had not even kept the Sabbath year as promised by Mr. Washington after the signing of the declaration of independence. As for how history has this tendency to repeat itself is in many fears and lifestyles. Here is a hint, what is the seasoning mentioned in the Christian bible more than ten times? King Solomon's temple was declared to be ran as how "right is not a fight", which is partially in a small way a teaching of life but in Old Jerusalem before the fire from the sky had been, because of the decree to honor the laws of which were Right and Just. Not what the Greeks had declared correct in their idols as "Might is Right". When only one family had been saved, they were told not to look back no matter what. The man's wife had been turned into a pillar of salt for doing so. George Washington is in one of Americas notorious paintings kneeling at a church making a decree to his God from which we as a newly made country will be for what is right. Little did George know that that exact same church would be the owners of the World Trade Center's land that had been destroyed on Nine Eleven, Two ThousandOne. However that church sustained little to no discernable damage worth mentioning, over a mile away was a building with its foundation in their basement cracked. The very same cycle as in Jerusalem is in today's hoe chain. It's said the first profession was that of a prostitute. No matter, that's only advanced itself to what we are governed by in every state in and out of this country. Commonwealths as well. There is no need to withhold that America has only 46 states and four commonwealths. Go ahead and look up the Commonwealth of Virginia University. How many readers know that there is a correct way to dispose of Americas flag? Every star and stripe must be cut out by hand and then burned. If there is a prayer in there or not it is unknown by myself, I had only been reminded of that factor of our flag by another Boy Scout such as I was.

For as randomly put together as what this is, it's not that far from the beaten path yet so far out in the distance because of its randomness that I'm declared different and set aside as an outcast yet I made my enemy my friend (enemy self to my friend-self) from such a lesson. Like a box of darkness is a gift, so is the lesson I learned for being set aside in education, and other areas of life and the pleasures it's not unleashed and over time aside in society as well, though I'm fully functioning with anxiety and the things of "problems". I can't speak for my generation but the person I am has not been bought out by games, toys, trinkets, and souvenirs. Factually not the first generation to have these computer-chip games, my generation was however the one that got the first good games. I will admit that I have had several Playstation 3's, just about every size made when the third generation had been released, and even a couple of X-Box's. I believe that I went through so many of them because I hadn't been listening. I possibly am now.

As the story goes, woman was the first to become intelligent from a forbidden fruit, then man. Woman's idea in my mind for any sort of relation to what I imagine, is the old way that had been used until the advancement that possibility itself had become. Life is so upside down that if you stop to think about life for awhile and try to realize that women rule the money of the world from bottom to the top, then the final relation to the aforementioned situation is to man being in control as the lie in the woven curtain. The very fabric we are accustomed for compliments, and even the fabric of time. Uh oh, now I'm in trouble right? No, actually in fact the contrary. Through the spirit of life, God had created Adam and

Eve, then they had betrayed him by the way of the fruit as it was a choice, though ANYTHING else was acceptable to eat. Hansel and Gretel, Joan of Arc, Romeo and Juliet, stories, movies, and everything down to the small things which play a critical role in life like our spirit are either our hero's or our brethren, and even our own perceived enemy, and a literal one. Spirit left alone will eventually be misplaced or as told "misguided" and even destroyed or broken. Some though, as alone can actually flourish. ENOUGH of the relations, the old ways of Analog to the new ways of Binary is not too far from the original spirit of life that had created life itself. Are we on a cycle from before and destined at one point to only be one in control of what the last instinct of life is (which is breath upon dying), that had already been created before? Creations maker had a perfect son, and that Creator eventually sent his only Son to explain the way around the heat of the oven, and the for-telling of the wrath that may come. Hell hath no fury like a woman's scorn. Thank you God, for moving that in the light that is life. Analog is portrayed in a singular wave if set at the correct tone, and can be altered for signal to a numerous amount of vibrations which can be perceivable or in an undertone. With a small amount of an ignition source, salt water is even flammable with analog wave technology. Nikola Tesla had found the way around his-self,(his own inner woman, because every fetus is apparently a female before the male chromosome is placed), thus the same in saying he found a way around the analog technology, but not the answer of what that was to be around those vibrations deemed analog. Edison with Tesla out of his name mentioned in textbooks for the majority of the time and Mr. Morgan had screwed him out of all things that could benefit the entire world,(though that from themselves may or may not have been their plan as to do so, or not)the government or those who have control, have those patent rights and will not let them go for even life. That may have been the Hail Mary man needed to create Binary code. Or was that theoretically implausible to say man is the binary or that he is the fabric? When necessity is desperate, the void must be filled. What has ever been invented from without necessity? Binary code is portrayed in ones and zeros. Either something has or has not something within it. Smaller than at the atomic size, and larger than quantum mechanics and computations is everything that the universe is made of. No debate on Sony's VAIO symbol of transition from one faze of life to the next. Not unlike Neanderthal from what had been his adversary and so many other mentionable places in history, a victor over a lesser thing. Mentally I know what many may think, and that is that man should build such a computer that it can store a human consciousness' as a whole species. If one why not every one of them? Is that the largest question out the gate? First let's figure out why and where the Mayan civilization had gone and disappeared to huh? Tesla was placed on a stagnant back burner for one reason and one reason only…for man to be the complete ruler of the place in life he had been created within, and the image of. To be the ultimate creator of the fabric around him, and alter or bend it any way desired, I'm seriously not trying to be sexist whatsoever. Not to have been placed second to what was made FOR him. From the air, the Mayan's, they have structural buildings in line with what we call binary technology yet they were primitive though they are not to leave a trace of their records behind for us to learn from them? From the sky they have built dirt mounds able to be recognizable as many things to be mentioned, only for this instance, that of the ant, or what we call infinity and the symbol they relate to one another. What it is that we need? Did they make a mistake or was it their improvements that had in fact been their history lost to us. Is that the key to life of humankind or the life humankind has in the universe? Well isn't that a nice ponder? I can go for days into places creepy crawlies will never dare enter and come out unscathed for the most part… all I have in my way is peace of mind and interruption (concentration), up to that thing called my inner self that is or was my enemy and at times still is, for confidence and to the interruptions that life itself can become, day to day through the mundane. First from within, and then from within to what isn't within, and not interrupting another spirit that is as well, still in life so they would not be interrupted, and "off" his or her concentration is how I would see the meditation theory for the missing Mayan civilization. But when was it bound to be interrupted? At the first point of contact with other intelligence that had recognized their conscious potential. Life itself is a genius and at the very same time is the exact opposite, or is it?.. Is it rather a genius and not the opposite but more to the larger side of what the stride of life is supposed to be? The genius and the leap of what stride it shall take as directed at the hands of men, leaping like so many of histories famous stories and portrayals that they become immortalized for the way they enter, deal with, and leave their comrades with peace of mind that things are now somehow better, life in numbers is a genius of its own societal drum. Saying that, it's becoming mans turn to be the complete director, not as though from the weave of the curtain of life or time, but as direction leads the world to a universal truth that it's possible to be larger than life itself and smaller altogether through newer and newer technology. But that's not totally right for teaching is it, the newer and newer part I mean. Eventually there is an end to all things good or bad.

I myself am looking for a good girl. That in itself is a troublesome opportunity and failure of possibilities waiting to become the avalanche of denial. A new life. Ashes to ashes and a new life from ashes that the new life shall return to in the new world over and over until it becomes obviously imminent that we CAN re-connect ourselves with the One who had created the universe, and even our past self. What then of an average person to envision what his/her future self and the future that *IS*? The statement existence simply exists is far too evident of a fairyland I'm not willing to weave myself into. What if that were true for reincarnation? Is a body that had left the soul any different than that soul leaving the body or is the soul even real or in fact a made up thing just to have the ability to take it away. Good questions yet I have no one to answer these things that I love to trouble myself with, for that would become another mundane attribute to distractions of my concentration I have not the time for because the telling of such thoughts can get one killed, or debated into stagnation and off course. I am and have been for awhile, above myself. If by this page you have not the sense to understand that statement then try researching introspection and then don't let education get in the way of your own learning. Shins get scraped as a child, and life being a genius itself makes that pain a little more shaded than what's expected from our infancy of innocence into our adulthood, no matter the kept innocence or forgot. I have not the exact knowledge of an instruction booklet to life and how to live, however, SO many people of either gender have created their own path, that I at a young age started to question silently and to myself the "what if" of life. Why not ask the large questions nobody has or will because it's too far from the accepted course man has been on? Otherwise if not trying this silently, I would be in straps and sedated to the point of a chemical coma to challenge the ways of normalcy like I have.

New chapter

Another gift from the unexpected conditions as the society's elite classes that dislike waves, continues as it has since before Jesus' time. His message was turned on His actions and made preference to what was at the time war, the beginning to the story of life as He was life's first teacher of that story. The right way to live, honorable way, the honest way, the leadership teaching leadership way.

Normalcy is only used in what has been the acceptance of the things that were allowed to pass as "ok" to onlookers or society as civilizations had begun and thus continued. Every day this normalcy changes and it makes me sick to think of how long it may take to correct it. There is however the quick fix to this from an average citizens mind yet still unheard. Others have their philosophy to life and how it's supposed to be for them. Entire civilizations have made philosophies around a single thing or even a set few things. What happens when the destruction is imminent, or already over, if at somewhere there had been a point of no return, a universality savior has spent his knowledge on faith to have all return to him if the rest of us fail and must be returned, becoming Déjà Vu. Right now, this weary second there are forces of man AND nature about to be beaten like an albino baby seal by the same exact forces that have excluded themselves from the ranks of society. If that is not who, then it is in fact the one true God, not intervening as to allow his creation to appreciate the thing it was made to appreciate as a whole, though it's not happened yet other than that of the Mayan civilization and the others who are deemed "Lost". The things man has done and as well as created, are destructive towards nature itself and changing too fast for nature to be as calm as it is. Living in the upper decks of their own yacht or below the troublesome melancholy seas never paying attention to the "Sailor's Devil" Davey Jones that they play a card game with, not knowing that they only have the one chip of history and it's already into the pot, the human race is potentially oblivious to the fact that there is salvation, but that salvation comes first from within. After that part of what is within gets to the mainstream of almost all cultures on earth, if not the most destructive ones first, then is the time that the advancement will be acceptable and able to flourish for all and not the few bastards that are as selfish as a spoiled child with toys not their own. Weather an addict admitting, or the bible teachings saying that it must come from your own self to ask for divine intervention, there is something unstable enough to admit this to today's world. Super computers never work hard enough and the mind over-rules them so frequently that a new code or virus can be or is written anew nearly every week or month sometimes. Quantum mechanics though, deals with the unfathomably small to the extraordinarily large and beyond. Remember math class? Here is the addition complexity boiled down to the bare minimums for anyone interested enough to support such an irregular task. (WARNING: This is about to destroy the earth by forcing the hand of a larger power who wants not to give up power they call control. Those who won't make the difference easy are those who return and make their selves their own victim of natural order, thus their own demise.) Not known to man in his mortality, if we fail at life, we mortals return to the time we were saved. Why? The story is definitely told as it should be in some places. This side of the story had been of Lucifer claiming to be better than God Himself and trying to do better works than Him, then defying God's orders to leave his creations Adam and Eve alone for God's favor that of human and not his Angels. What that is for today is mankind spending entirely too much attention to the speed in which it can advance itself. SLOW DOWN! Making nature bend to man's will from machines to control weather is wrong. Making weather patterns completely re-arrange their-self from trying to divert large amounts of warm water farther north on the west coast which inevitably caused hurricane Sandy (I had verbally predicted thus entirely before the year it had happened, as well as an entirely longer than necessary phone conversation to whoever they were, though have done nothing to reward my excellence to their lives.) but now more recently how Buffalo NY had been buried under seven feet of snow. These things will not stop any time in our lifetimes unless the problem is removed. Besides that factor the H.A.A.R.P satellite creating patterns of weather change, that disrupts the nature of simple order in nature still faster than nature can keep up is wrong to do as well. On another note the chemical trails that streak across the sky from horizon to horizon dropping poisonous amounts of death, where as a child those streaks had never been there, I don't know what to think of it. I look to the sky's now and then but the horizon and I don't even remember a day when the horizon had been completely blue as was when I was a young boy, because it has been so long, possibly in my previous life I recall this blue color no longer available on the planet. This warning is to reattribute what wrath can be, if ignored. History has itself not been silent and words here are not empty.

Goal: quantum computers to deal with the infinitely small to the infinitely large, give or take the infinite part. Add that goal like this, nanotechnology, 3-D printing, adaptability (the tricky part) and free technology like it had nearly been when windows had made its revolution like Fords had. Such a radically new technology for nearly pennies compared to the idea it had been. Free technology is Nikola Tesla and John F. Kennedy's interest free loans smashed together like crash test dummies but in symphonic harmony with seamless universality. It's the steps mankind will have to take that will insure itself the survivor of fate and evolution and the God complex all in one and the no death scenario has begun if such leaps of vigor are undertaken until completion without today's interference, starting with the majority of crude oil left behind. You haven't heard the kicker yet though, which is my own as I believe it to be. The kicker is to re-write math for what it currently is not. As for experts to discredit this is not entirely out of the question, however Einstein was discredited for many years. This new math is like this: one plus one equals out to be forty eight. What the next question is to many others, is what two plus two becomes, which is not even close to ninety six, rather the question is what is 1.1 plus 1.1?or even if through these styles of analogy's you can count to a new understanding of ten, can you add things to make greater in the end? The things that are possible today are decreasing the need or even the desire for war even more nowadays than ever before(congratulations…now see it, then use that knowledge). When A and B no longer add up to be C, and was addressed correctly no matter the way, is that the first math question of all time because several answers exist for the realm it would become, possibly beacon to another realm? Mans understanding of himself will change if this is perused and debated enough to bring completion of said debate…to create their own first universality to the universe from which it has not yet been done here on a large scale minus my theory of a missing culture or two. Earth this very day has enough space, food, room for growth, and exploration to last four hundred more lifetimes after my own. The resources however are not infinite. Those that are finite in this world are namely fossil fuels that cause the wars of and for power because someone once said this land is mine and therefore better than yours like two children on the same beach building sandcastles. Put your dicks away "Gentlemen" and grow a pair, be the bigger part that life demands and not what man desires before the war is literally over the water itself on the whole earth. Here is how to cease mostly all wars that have been over all the oil and the foreseeable future over the same thing. Neodymium magnetic material is above and beyond the call of duty if applied to transportation large and small, including air travel. The manmade material is the ability to change the world without leaving powerful people completely powerless, but within their lifetimes, un-noticeable to the rate of decline. Has it ever occurred to anyone to have power is only to have control over their own self? Anything beyond that is too much to comprehend within and outside self. Anybody who has ever had children can tell you that, but here I am telling the world this with only the hope it will continue the way it's intended to be, not how God "Left It To Man". God forgive me for leading astray those who won't understand re-direction. Great things are upon this day's age and if we as a people are too blind to add up the factors and risks from so many pools of knowledge, then I'm not sorry for the failure of their own blindness because many have tried to show the light only to be shoved stage left and burnt out or killed and or shunned by the blind lead by the blind, lost to history.

New chapter; A Rant?

Adapt or die. A lighter to a caveman wouldn't have given him anything but a fat belly and death through lazy attitudes towards less exciting accomplishments from himself. Fate either laughs at us or smiles upon us with gifts and blesses us with joy or sadness. I hear that grey hair is that of a crown of glory earned through a righteous life. If you don't fight for what you want now, how do you ever expect to get what you want, or appreciate, or get anywhere, or to accomplish that thing humanity needs, hand to your children or grandchildren that which could have saved the world? Overall good is good, and humanity as more than an acceptability is too far from a complete control over reincarnation and whereas to put their next of kin or self after death. If you're given something that you once desired to acquire for yourself through hard work and appreciation to get that thing, how in earth will you appreciate it as only an answer to the gift someone could have just handed you? Nuclear decision time, not only the Japan nuclear disaster cleanup, not only the nuclear warheads and atomic bombs, but also the nuclear waste dump that's collecting mass quantities of garbage floating out in the middle of the ocean and can be seen FROM SPACE! More than what a war with death is, war is also in other forms as well. I'll explain, a five point war currently consumes America with it only having fifteen percent of the world's population and fifty percent of the world's military sales. Another point to make is that from those exact leadings comes large countries such as Russia, and China to whom America is in debt to for God knows how long. Russia flexing oil control in the east pressuring America through the United Nations for intervention, almost the entire middle east in conflict, all whom fairly have a bad taste in their minds for Americans, and another notable idea is the supposed war on drugs from what is within as to what is also on the outside. Then some ass in government put two things together like he's a genius or something and said there's a war on terror itself…never ending the conflicts we create, and North Korea staring at us in disgust for our ways. Who is to lead you if you will not lead yourself? An entire people on this planet has been rounded up for the wrong thing, and it's the gas chamber through the push of a button in a suitcase that'll obliterate the PLANET eight and a half times over its own destruction capabilities and growing. And to make matters worse than that, not only is this suitcase in one man's capability, but many. What a headache huh? To the disaster though of Fukashima that has radiated much of the planets oceans, (give it time) there is a multitude of ways to clean the waters without destruction. The most obvious one is an organic process that uses hemp to grow on the ocean in fields one hundred miles by one hundred miles square. Over one hundred uses of this material are a possibility for the plant itself reduces and expels the said radiation to a less powerful radiation. Start with one of these squares if the politics deem it too costly (fools), then question the amount needed for a quickie fix. Then Someone will have to have the courage to stand up for the necessity that it is, and begin to either pay for, or collect donations used entirely for said project and not pennies on the dollar like most donation projects used these days. International waters have no jurisdiction in any country, so be careful about the protection of such a thing. Hemp can grow in many conditions and is also able to clean radiation from dirt, seawater and even the air at the same time. Daily use of safe hemp after cleaning is more cost effective than anything that I've heard proposed thus far. Or are those who claim power not going to do ANYTHING? Tri-fecta in-between the developing Eastern countries and their pollution and the Americas which has their own oceanic problems to correct before nature and the levy break to disrupt all life on the entire plan of life, which mind you is a player itself called nature, not needing to be the physically seen thing most believe a "thing" must be. That's enough of a spread to make a trial run worth the time and effort for tolerance of natures backlash(though many swaths are needed to make a fast amends to mans much needed apology to earth who is in fact not our friend). Now that the backlash of nature has been brought up, what say you of a lever long enough to move the earth? Not one can be built but many can be made to do the similarity as of the one imagined. Windmills are yet another one of mans creations I'm proud to tell you what nature is unhappy about because its making her lose her step ten times faster than you'd expect if you were an educated scientist not funded by government money and tied by government tape. She cannot keep up the pace man has pushed her to and is stumbling in many areas. Signs are everywhere if you only listen, or open your eyes, or pay attention to the wind for example, and undistracted. The accommodations to slow the earth's rotation is far too much a feat to be proud of, therefore I'm dismissing any and all honor for the accomplishment that the Three Gorges Dam has made for those who have had any, and the country that had made it a project of accomplishment rather than that of a simpler innovation. Here is a thought, the first lightbulb was a gem with two drilled holes and set at separate angles with metal rods inserted for the power of friction to emit a light for the purpose of looking into the darkness from the spark of friction(millions of years ago). If that were the ease of the used in and of itself, today's problems, would not be in existence as of yet, if ever. Then not one word would have been said by this lonely man in his late twenties as his body shows. Value is of itself and useless as it had been thought of from the first time to whichever the last may be. What one deems valuable is not always the same as another's view, thus a dispute can cover vast quantities of problems in short amounts of time. What is pretty? I'll ask in another more definitive way, which rock is more delightful to look at, the one with a few stripes or the one with a few stripes and a few dots? Either and both are nearly the same color, but one could be considered more valued why? So short this time that it disarrays those who are merely by-standards willing to live to their peaceful nature and one with the universe. Not even through drugs can such a connection to the universe be found as those to whom it has explained itself, unraveled itself, and shown itself to, though both ideals are accepted to some degree. The few that this has been done to, they will always be held accountable like a man with a button, though not as dramatic for they have no move the universe can't dispute or reclaim its own after it's revealing itself to them.

$1 + 1 = 48$. Why? Baseball strike out rules of three and out, when meeting people for the first time or not. Both individually, and as a pair. Its inspirational, has justification, rationalism and is best for self and the possibility to a group for the equation to become the equality race to or from 50 to whatever perceived goal. Give a man a fish and he eats for a day, though if you teach a man to fish, he can eat for a lifetime. To continue the search for more, and what is greater than what is already solid in our minds is philosophy. $1.1+1.1=?X?$ Pose an odd question the right way and an answer not uncommonly phrased may be the answer. A real world life formulation to what is available not only through the mind but the minds of others as well is a great leap forward. Life and all within life has a maturity maximum unless pushed to the limit and what do human minds have a limit to besides their own self?

Tyrants are those who suppress and create the situations in which suppression continues. My goodness, I'm threatened by this, and will not stand aside. Things get even more emotionally heavy as our current leaders argue a point to disarm the citizens, and already have in place a "law" available to be readily called upon called "Martial Law". From protecting the life they have and the lives they are looking after, the citizens have human rights as any human. I myself do not support the supposed need for or the use of any such law against the people in any country. It's already threatening me and will continue until the threat of force is removed. This frail line of life we are bordering is known as rapture, Armageddon, mass genocide, or if you wish to get serious enough, the destruction of the entire planet and anything life sustainable throughout. This fragile leash our political leaders are working with has been over time from our founding fathers of the country we stole, is so fragile that a sneeze could break such a tiny thing. This is not bearable fruit from the harvest life should sustain and has not been doing very well.

We mistakenly believe the citizens are fully capable to vote away the leaderships organized wealth no matter the cost to their protection of it. The whole system which laws are built upon is crumbling and they over time have timed this to a particular advantage in their favor. Decades and even centuries ago this had been started and worked with. The opposite side of this factual lifestyle we have been suppressed into, though disbelieve because it has no "immediate" threat to us or our children is that if arranged in a greater universal way, life can be profit driven for REAL. Profits are described as a gain to oneself, group, and society as well as those in need. Those profits are gained and then equally shared and considered for implementation and for the group, from the group. More on profit to the community later.

The honor reasonably taken from the project which had slowed the rotation of the earth by man's hand is possibly a temporary thing to deny them of having. A feat of engineering yes, impatient even more than a feat however. This project had finished ahead of schedule in 2011 or so but now in 2014 the solar capabilities are far beyond that which was the cost this said project had been, and even before they had started to enlarge the electric output, the solar capabilities were already acute for success. And relocating hundred year old homesteads for progress is unbelievable. One house then could sell unused electricity to the city in which it lived, but the discharge in the areas where a cable like that to the nearest town would have been exhausting to upkeep. Therefore, provide these as the roads have been, and eventually the use of solar and magnetic roads will be in place before the grand-children's children graduate high-school. Logically the dam was a foolish project unless you consider the depth that the river had been prior the project. Still cheaper other ways I believe, though it's not where I live, only it's where I owe money. Unfortunately because of my "Leaders" calling the banking system foolproof to line their own pockets, babies in 2013 had been born with an average of $50,000 in debt or more, with which we won't even account to the billing from the hospital and their services. Nope, even if a child was born at their home, in a car, or a bath tub, the child was a debt to society and that is why I believe the politicians are in fear of their own lives. Like back in the book of Daniel and the ever growing population, the leader then had fear from the numbers he could not grasp. Fear of losing control is the greed sin is capable of. Genocide is but a step away sometimes. No sugarcoating this, it's a possibility to many places on this earth if not the entire face of it because of the D.U.M.B military bases deep within the crust of the earth able to sustain life for many generations if not all time.

    The good days of yesteryear have been set aside until these matters and more are finished to God's acceptance, or this world dissipates into oblivion from its own deeds. I myself PROMISE that the noises to which were from unknown sources all around the world in various areas, was in fact the heavens speaking through the same music we people have destroyed the heart of, within ourselves. We are no longer in tune with mother or father and it's become far from reach if none are to stand against wrongs. Though a creation of man could have been the reason from space skimming along the atmosphere, yet, why chance what has been told through the generations like that? I refuse to bow to anything I could dispute as mans work. God has called names, and when God speaks it's unlike anything a person could have related another experience to. That same vigorous experience is grace in all things God does. I pray this world tells its tale, no matter the cost of correction God Almighty has set upon it.

    Which troubles me more I have not the slightest clue for how my heart feels, but my quarrel is this: the stories of the bible were recorded and told because the time of Christ himself, but what's been as of since? The other side of this question is that of what is as of recent? Why is neither taught to the people who still hold breath, and for those who may not receive such a blessing? Is God's work done? No I believe it not to be. Are the important things no longer to be set in story like it has been so? Where are his teachings taught besides in a singular day andeven at that for but only a few measly hours of worship? There is something in the way and God refuses to remove it for man to appreciate man's own possibility at success.

Either that or THAT IS the answer to what it is we as a population need. The fights over such answers have cost much while the world changes and we change the world. Nature is not an onlooker oblivious to what have we mortals changed. A saying said as though it's better to catch flies with honey than vinegar; however the use of shit for the catch is the probability of success more often. So the stage has been set, and the commonality of which fertile ground is made, has been simplicity at its finest for God and his game played with man. Seeing is believing and now that man is not much of a challenge to control, the lies and deceit will baffle those who try to live in harmony blindfolded around those who also attempt their lives of just cause or not. Education is no longer fun unless you force yourself to conform to using deceit upon yourself that things are alright as they are. Learning has been a deceitful dread to the continuous new generations run by those who chose not to lift their blindfolds to find a passion of good deeds.

Let us return to the definition of pragmatism and place a relation of slavery to when African slaves had to purchase their tools, clothing, and daily needs from the plantation owner. Not to mention the mining towns that used company money and company funded stores where any other money was useless. Slavery again has been in many places not only on American soil, but it is still here today. The banks got the bail out when if they hadn't, and instead the average person above the age of fifty had been given one million dollars tax free to use how they wanted after three stipulations had been met, this country would be dancing and would have started dancing by the time the ink had dried. Those stipulations of receiving this amount of tax free money is that their mortgage is to be paid off in full, they must purchase at least one brand new American made vehicle, and for every child they have, they must put away $40,000 in a college trust fund. Any and all other finances thereafter can be used as they would have pleased, including and not limited to the stock market. This form of debt release would at that time of average citizens age of over fifty years old by sheer numbers would have been less costly than a trickle-down effect as the lie we were sold by President Bush and then again by resident Barak Obama. If done then it would have been greater for the economy of our nation, thus boosting the interest of the American dream being chased. The politicians know and knew then that that was what was going to happen because they LIVE to do such demoralizing actions against others. Those citizens over that age are more responsible than that of those who are in their later thirties for sure. Retirement completed, and no more social security needs, effectively reducing the taxes collected from the rest of the working classes, who take the brute blunts of which fortune 500 companies pay not one dime in taxes, yet control all the votes. Today, corporate America owns our homes, the education our children receive, water we consume, and everything else which creates a deeper debt from overseas and from within consuming everything from candy to jobs and car parts made in America…, and claims the greater good for doing so because of their innovations that they set the price of. Only so that we are in a constant state of need of them to continue suffocating our lives by technology or social goods fed as lies to bind us to them. Blindly though we act as though we do not need this and can easily turn away their social classification as blue collar against the white collar. Blue slavery and white slavery is how I could call it. Electricity could today be free and could have been free since before my time and my parents' times. Chemicals in our water supplies only to justify a "purification system", when all it is, is the ready button for social cleansing to be "valued" and "orderly" or even to the point of "completely accepted." And for the explanation to those losing their blindfolds, social cleansing is not washing your body in the river, social cleansing is a river of blood to restructure belief systems accepted from force in a hurry. Born into crushing debt, no hope of survival. Deeper into their grasp of slavery. Independence is a human right. Inequality is not universal. Civilization has for the benefit of itself succeeded in the past on other planets by now so why can't we here? What point of no return had they gone through that had their planet(s) system change into success and not stagnation? What blocks our path in a large scope of reality and great responsibility?

Exponential expansion is a mathematical nightmare for what pi is to its own end. Reference to cancer as astrology for the stars around us in picture, pros by symbol. One in the same in reality, but one the spider and one the fly. To what mathematic evaluation is from a numbers stance is what boggles my mind to some extent of insanity, so therefore I decide not to use much more than I need. Computers can handle the rest, yet the education board is not up to speed, the politicians cutting every other year, pay-grades are inappropriate and as an extreme example of an in your face look at how wrong a "super-power" has become to its own people, in some other countries, the students get paid a handsome sum of money just to go to school. Music can no longer be as malicious as it is. Teaching students what a natural tune versus a tone and or a note is an extreme impact on society that hits the youth ten times harder and quicker than any politician has in decades. Idols worshiped for doing the wrong things and nothing but more pay for the first pictures of someone to be in this week's tabloids where it's an exclusive every other second. Now, I'm not for one to wish in one hand and shit into the other, hoping that the wish comes first to save me from the shit, but this shit has got to stop. Education is not up to date, regulations are getting out of control and into the daily lives of every day children, and businesses can't develop enough to grow and thrive because this country is failing, and so too the world.

How easy is it to confuse pity with mercy? There is a realm in reality yet constantly a non-reality that bores those who look for it, and kills those who find the way there by fear of it being stumbled upon. Innocence like any other place imaginable is as real there as where here is, but how the innocence is handled can vary to such great degrees, a giant here would be but an ant in comparison with humility for its handlers ways.

Listen to anything around you at any given moment and even at its most natural element, there is the insanity to it. Nature is un-natural and equal to equality but as all things are subject to disarray and drastic changes, perceptions and personalities get the exact reverse reverence in the warmth of acceptance. Such a repertory exists with, for, and because of the universe that not all existence is naturally in our view as decent or justly to one another. Reverence is a feeling of deep respect, mixed with wonder, fear, and love. We as all notions of conscious life in any galaxy will and have a reverence to men of noble lives. What acceptance here is noble to the majority of man because of history and the universal goal expelled as truth, though it's been shit. How that disarray begins wherever to destroy or excel the society is completely in the hands of the people and their tolerance to what is the goings on around them.

Rally ye sick men and women and children who feel compelled to come together for a common purpose. To heal a sickly man, to a sickly nation, and their planet as well, what is the accepted way to say goodbye to He of whom takes sacrifice in His own name for the greater good and who will behold Him in great reverent speech and truth? Not killing in the name of, as the Christian crusades have done to rally by force and threat of force or death by many means, that is not as how justification is seen from the first teacher of the story of life. But to fall and not be known is like the tree making noise in the forest as it falls when there is not a soul to hear such calamity. A wonder would come to be more than He when reason would allow itself to be uncovered by mythology getting created by rumors and deceit. A rumor I've read is that a painting exists of Jesus' shadow that of a crucifixion represented his death for mankind beforehand. I myself have made my own painting that can be used in as many ways as permitted by the community as they should have the complete voting power for their survival needs, never to be taken away, but as a way to see or learn more from many sides, not just the one sided facet life gives of the strong over the supposed weak. Modernization of mans evolution that technology has brought him is allowing the revelations to begin to unfold in a favor of good or evil not yet told, though still unfolding and unstoppable. Time will stop for no man, and slow for but our God to make things His own in our own end. Preclude paranoia, shut it out like a dream unable to be recalled. It won't matter for long, for man is in responsibility to all things with life. Pound foolish and penny pinch is no way for your leaders to begin a new century that shapes the sky of what dreams may fly within, or die without when so many pools of unused knowledge are available to common sense, becoming every day livelihoods of all ages and those who have to say farewell before a way is found to live forever. Naturally it is done in the story of life when it includes the use of time and legends but for it to actually become a reality to the body, nears this day as Jesus fell for you and also because of you, by the design of guilt told to you by the leaders who now tell the story to you for the complete reverence you hand over not knowing it's a fine silver with some apples of gold mixed within every now and then for irrationality in all its essence and glory for good and evil ways you know not of in most definitions called a reality to life that should be available at the ever younger ages so someday a newborn baby will speak the moment it breathes air and for clarification I put my life on it that thus can be done yet that ship so far is not coming back so what are we waiting for? Won't a newborn talking be the newest glory of man's ability? What if it had been done to the technology though it were a natural every-day event no further than reality is to gravity being the most unrealized justification to space and all things within it? That, my fellow men of honesty, is true glory through kindness. It must be known it won't be in one or two tried generations but more like if diligent to the aspect of patience and endurance with tried and true techniques, then in or around ten generations in mans capabilities. Greed is almost always going to be in the way when the money is involved, like a gravity to a planet, greed to the grave. That it be like a story learned and told through the breath of life carried to the child in different forms at different times.

Here is a tale, call it a fable if you must. At a point of choice, one can look back upon pre birth and murder by success into the egg? Waving goodbye or making promises as the rest die off, for the rest of those who have not made it don't go away, you do, and some as myself had made promises to those which were nice and as accepting of loss as most everyone, but the anger from some is almost an empty promise of revenge if you were to believe it to be the ripple as it's said in the fabric of time. What you carry with you is more important sometimes that what you don't. When a person introduces another into the universe, at what point is the mentality jump of the universe a natural occurrence to the benefit of the universe itself and not just the planet the life is on? Even mature children with their imaginations can perceive and conceive not only the plans to go forth with but visa versa on the imaginative maturity. That is the point in where innocence is by you either preserved, or denied from you for the rest of acceptance, until rediscovered in time and dealt with accordingly. Understanding that consciousness brought to itself, can create, is in the lead for those who will be lead, could be preservation or perseverance. Nature versus nurture is on its way in this story form so just hold on tight, for this is a ride with more than tide, and there is now no stop desired from creative honesty no matter the age. I feel that that is right in itself as so I will go on to do so soon. Still, let alone a choice with no adversary to which is Alpha and which animal, basic needs no matter the life form in question must have its own basic natural environment or adapt to survive. The deceit is that {the fight wins in itself if, the question or action is not asked or acted upon}. To massively speak at birth, its reasoning must be handed down with the utmost and greatest care. Immortality can come to a/the shared understanding as the word is, but to be the practice of a better man (the noble) pursued, perceived and put forth, patience (from a nobility) within all sides is first understood, accepted, and then planned out, handed down through the ages to the younger and younger peoples for the advantage of speaking at birth with the bread of innocence yet within them.

Please follow me on this path for just a few moments. Surely a wise man can be asked a similar question as from a stupid person and not learn from it, as the same type can be asked from a(wise) stupid man to a(stupid) wise person who may learn. But if a wise man can learn something from a stupid person yet a stupid person cannot learn from a wise answer then how far has earth's stagnation been from the thick of shit? Even though that is wrought 'full of itself' as it's typed and not understood, it makes sense to say as it has been this situation. Life is nothing if you and I cannot fathom what may and or may not be, at the same instant it is or is not a constant abnormality simultaneously disassembling and reassembling continuously from something to nothing all the while that instant had been just that, an instant, and then on to the next. To share is to lead in all proportions, one of the best ways to lead. Mathematically speaking but answering in a question, what is the smallest measurement mathematics can fathom? A thing of infinite angles and any smaller is non-existent? The re-creation of it and breakage point is matter in attempt to be in one realm or another, and as I'm psychotic, there is a door from one to the next, the quest and question is how for both quest, question, and the person who is on it (yes both is in that way three said extremely abstract). It also in the other way of spectrum measurement or theory that of a black hole forming from nothing to existence, and even to that theory of dark matter which wants to not exist at all for how its explained from what I can tell. Too much time is wasted on the particle colliders and things of the like. What happened to the sciences when they were brought out by collateral constipation of thought pushing forward society? Droughty greedy corporations and a select few who can fund such a waste of time and sciences efforts stalemate progress. Science has nearly died in my non-professional opinion.

Is that ever in trusting awareness that faith is not to be as a bloodshed for any gain unless it's for the truthfully good, meaning protection and the protection of the threat of force, from within or from outside forces? Centralized banking is an internally driven factual force of terrorism in everyday lives, masked by it having the definition of "necessity" when in all reality it is not at all. Any farther is too destructive into daily average citizens' lives. Reason will permit the skip to this if you see the flaw because one; the flaw is later found and fixed so set this flaw aside for now, and reason number two; it is the trending of what today is called becoming by right an adult in many cultures and even practiced by law but not at the age of seventeen, the debt owed to life itself. Because seventeen is only one year younger than eighteen, an individual is already looking into what they will, ultimately upon availability, be doing if not already on the path to creating those exact circumstances for it to happen if that person had missed this as they were once a child. That in effect is the multi purposeful base model for the next thing within the addition to authentic life of what is valued interaction in your future if additionally explained in abstract modality. In other words, the innocence that is within an individual has not shortened in times of evolution and technology, but actually gained onto it the older ages when its maturity is focused on, and then if disregarded, lost for awhile. Its mans own responsibility and works that had brought out this good, and now that it is needed recognition, the responsibility changes to mans next item of desire and prayer in faith and WORD OF HIS POWER.

If in the first quarter of any given life of human genetics a person is to look back and say, that took far too long to get to at the age of roughly twenty five, then understand that as you get older the ages are repeatedly told to others as slipping by, too fast, where it went they have not known, in a type of fluctuation as this math undermines itself in abstract ways, it can be attributed to life even more than it had begun as one plus one explained in the first sections. This form of math is not sequential order but instead, lifelong and unedited towards the desires and prayers, and therefore the look upon the accomplished things one or another has done and to what extent. To some this is difficult and to others this is completely understood and they feel as though there is no other need to follow along with the majority that has been before them, but the desire to teach this powerful thing is there and I feel it within my entire heart when I look for but an instant, any longer than an instant I feel its fragility is too far exposed. Feeling a little hairy here I don't doubt their confidence as yet they're only used as yet in their first stage of life and to be entering the second, however their confidence is like an uncontrolled fart untrustworthy of any a-typical sense in an enclosed area and among strangers, as dangerous as it is for a teen to squeak their voice uncontrollably and without warning and often humiliating. What the words can be are as just as dangerous to those around you. Not to demoralize the nation's use of a good laugh but is that our response to the rest of our lives, for it to be as misbehaved as what should not be trusted as neither is flatulence? If at the other end of life looking back from an old age (and possibly bed ridden I fart uncontrollably, it's taken into account that it's from an uncontrollable use of muscle tissue and therefore a sign of decay), I can see where I had gone wrong and given the choices to change their outcomes at many intervals, would I choose this or would I decline and recline into death? That depends on the surroundings of the known world and how it is lived through others' lives portrayed onto your own and how you felt the first time you lived it through, and passed it through expressions of faith through one another. Had you been in a happy living, you would change it in a fewer multitude than that of an unhappy lifestyle, and had you been upset at a felt failure was every turn, you would most definitely choose to decline and recline. But yet here we are back to the scale in which this abrasive abstract mathematics is undergoing operations control from the front to the back and the back to the front being almost unrecognizable, although very similar indeed, needed as a new addition to taught men for understanding knowledge through question and reason, not forgetting truth in God's word. When will we get there is when we can look back to where we had come from. Additionally to the life you live, is the way in how you treat your kindnesses towards another's and how it makes you feel emotionally inside yourself. Emotions feel controllable, yet uncontrollable at the same time and to some, the very same instant is the both of these at the same time and extremely elusive as to why or how to reconnect the necessary dots that allow a conscious connection to what we want to feel more consistently. Only through the lack of knowledge this is evident, and even God recognized saying "My people are destroyed(cut off) for lack of knowledge...." Lack of knowledge has caused men to be destroyed and lives to suffer needless loss. Even if you're not diagnosable by science disorderly or mentally inept, we all have felt as though a dot or two (marbles) are missing and need replacement at some point in our lifetime. One guy says some lucky bastard is out there having a heart attack, and the heart attack victim is saying that he would much rather go through a bone marrow transplant. In either of these cases the emotions of the connected dots are socially to the individual person the same acceptable level. If however in the other shoe, and then the same outcome as that shoe provides is that of what is experienced, a third perspective can be found? This is through the power of correct prayer, not self defeating "I'm lost" prayer.

If without a crutch and in need of one to walk or move onward, you can't find one or be given one, necessity of survival at risk in this case dictates that another solution be found. Therefore when from seventeen looking to death, you see struggle all among you and all around you that fate is deafening to the heart that gives us emotion, act as if you are as Jesus had been your teacher and you will begin to hear your heart and the gifts he has set there within all men. Legislation and regulation are deathly disruption and in everyday lives, from the cashier to the executive of a small company trying to grow. This number three added to one is of the utmost important to the standard in which we uphold ourselves and the confidence in which we may dare tell our self the truth, or not dare and lie to our self to go flowing deeper into the abyss like the rest. I will not submit my resignation to the fight life is yet, and even like a bullet to the head some will live on and so shall you if this is yours to behold through actions taken into honesty of knowing thyself through thy heart to know thy mind better. If thy heart is not with thee, then a mistake has to be corrected or mended for the bridge to get you where? To where is the destination the kingdom and wisdom of heaven through creation is. Personal judgment is as the Bible says not to be understood to mankind for it is far too deep in and of itself and defeating to attempt. I would agree but if you cannot control the emotion, you cannot guide the future, if you cannot guide the future you care to take, then you have not the choice and for no choice, life is meaningless and this cycle repeats, and repeats. Reference, Matt.12:35.

This thing of nirvana is elusive and as well as that is, also it's a sort of lie although it's a real thing.

I feel to live in a comparison to another is as much of an insult to yourself that you can give as possible because if even to an achieved athlete, a martial artist, a great person in essence, is not you and therefore, you are only defeating yourself by not attempting to be what you yourself were meant to do and that is to take on the responsibility life gives, shoves at you, and thrusts upon you at times even when you're face down in the dirt. Now to clear up what some consider a "perfect" time to take responsibility into your own hands is irrefutably wrong. Calm waters run deep and if you know not how much patience it takes for a man to be made through patience alone, then you won't understand until you're ready, that not all people are as you wish them to be. Immaturity exists and it can be crummy or it can be your defeated adversary. Let's talk defeat and bring into the word a square. Quantum mechanics deem a cycle similar to a square in relevance to time, continuous. Look into this for yourself and see that time is always repeating, constantly from the beginning and through the cycle then back, and from one point never to return to its original state or form of time. Though I feel there is a fourth which is abstract for the way it's directed into your own fashion, the time you have within life and the occurrence that YOU are YOU. The square is four corners, but cubed is six squares made into a closed box then within connecting the dots you have a diagram of whatever the definition may be. Taken three 90 degree angles added together is 270 degrees, I say to you is that not every degree of a perfect pyramid but only one complete corner lacking the fourth for the cured squared item of an abundance? Find yourself the references in the bible and you will understand this better than I can explain myself at the current time. I myself have a sense of self that desires to know the larger and more unknown to life so I have my corners set specifically to what I desire to find. Each completed corner in thought is given an understanding as are the other three of any connected face, then again connected to another three faces of other squares for each face to represent a whole in itself then somewhat if not in perfect maturity the completed cube after some time and more effort. Then added onto another connection of what is around itself as the cube effects the things around its origin, the vibrations not only words can be detected. Seven dots can connect them with one dot in the middle for your inner self to what is "out there" and as well as what that space is, being what could be at the same time or not yourself as to, for, or from yourself (prayer/control over self). Even if through yourself you digest the means to consider thinking before speaking through heart and a patient tongue, then you understand this consideration though you may not have the words to complete the corners, and thus your conscious connection to what is you and around yourself, possibly your greater unknown works. Whew that's a hard thing to type out, even if not to understand it from never thinking of it before. I'm debating on showing what diagram I consider my own definition but I have my own problems and have created my own mess so why send it out like a missile to destroy your own potential. The never ending basic untold definition you diagram of a cube six sided with seven dots connecting the sides from within, to the center of the face of each face is a general outlook to what paradox you find interesting at any level to question or have the understanding of a "thing" for many purposes. Each angle of each square has a meaning or definition or equation, signifying that it has substance. In between these 90 degree angles are the degrees in which you make up the smaller movements and if the consideration of the cube is at a large size then the larger more universal sized. What deems the frame a named or defined action or definition of action desired? You will decide from your heart what it is you desire, therefore you are the creator of what you want in the end of this cubit creation.

So moving onto the greater point of which is the first larger point brought up in this context is the way in which the cube can be deemed eight points. It itself or a whole can be seen as two ways if not more but basically I only see two if not looking at the connections life is of everything life is and you within it. Only let it currently be contemplation until fully trusted and proven. So eight points, seven of which are dots connecting the faces of thought bringing into the picture why a cut of cloth can be greater than that of another, when in reference to a person or another thing. Back to the butterfly effect and tsunami potential around the other side of the world in analogy, analog waves are that in which first God had made the greatest angels capable to move around with ease and later considered the fabric of life from what He (God) had made available, the light that gravity can be proven through science be bent by a planet. A cut above the rest is a great type of confession to a person when speaking of them to another in high regards. They speak of cloth, and the analog wave is but one string of a "quartet" larger than what life may understand it is. Abstract was normality in a sense at one point in time and yet still is relative when in reference to the many ways man still uses it and not fully understands the possibilities it has. Definition in this way is of no use but is usable if even for a moment that it is needed to be the abstract it is at that moment if not even this instance. Anyone following want to try and use gratitude? Thank God then. All that this is, is within all we think, see, feel, touch, express, and invite in, and interact with as it is of a useful tool for a greater thing than what ourselves are (faith), and have become since time was expressed unto us from before us.

In effect we have now just counted from one, to two, to three, and then to eight. Why has it been to eight might be an inquiry to ponder another way, though I feel that if you take the given three from what had been found as eight and you get five, and only THEN ask why, it is because of the deceit you tell yourself of what you CANNOT do because you are not due of it, though you are mostly made of the similar dew that is in the summer mornings from the mist dawn had brought. **God spoke "…Let there be firmament in the midst of the waters, and let it divide the waters from the waters" (gen. 1:6).** As the Bible says this is the devil in his betrayal we now get to exclude six and the seventh is already somewhat explained and is known as Gods perfect number, we, not forgetting the number system counting to ten have only two numbers to touch left besides four then five. Nine and ten become the reverie if you make it so. This subtraction may be deceit or the faith you live by because it has been set correct before you and all you have to do is faithfully follow what has been in as much a way the lives before you as the lives that have been in a future sense not yet created because this revolving door is as of yet to be surpassed in mans guidance and control. Prayer can be a cure and can also be defeating in as how it is asked for. Therefore when in watching over the things they were made to take care of after this said thing had been found and done, the cycle is returning to itself again and again until the cycle is broken. This is the reason this is my works admitted into known things twice over, once in the beginning and once in the end of each cycle. Now is the reason you must make it from yourself to yourself for yourself and the others that will follow in time what you do, as you do unto others, not blocking or holding against them bad deeds in faith they will fail, for what you ask, you shall receive. An eye for an eye, and a tooth for a tooth. Creating tomorrow better than today or the cycle continues as judgment is given as the end. For with God nothing shall be impossible. Heb. 1:3 (He is) **"upholding all things by the Word of His Power…"** That's said exactly as the Holy Ghost had intended for its correctness. Its intent had been and still is as it should be, the way He wanted. The word of his power makes these pages relevant to life and its future from what should not be nearing failure of our own hands.

So as to get to the dreadful economy, what it cries for, and the people who "control" society in the masses, towards a debt consuming tomorrow and what I may leave behind for generations, careful consideration with no levy, must be something like one to another when one season is not the same when and for afar, to another season of the opposite in range yet within scale, enormous. Problems exist and there is not a single relic on any life offering planet that can or will fix them all. Possibly in part or whole, belief is belief. Acceptance from myself to another is in that way nullified in such ways as why a candle would be wasted by burning at both ends or not if utilized metaphorically. If a result of friendship and or acceptance is not given, then it can boggle the mind as to the workings around the need versus a desire of having it. The question suffices as many as many may be while the answer drives the pursuant to make greatness greater. Adding to Nirvana is absofuckinglootly possible. As for what continues us on journeys as you could use a post to lean on, momentarily pause and inhale, make a count to three, exhale and re-analyze, re-adjust as necessary to begin the betterment of society's acceptability one person could create alone through patience for the rest to create from such a thing as the masses massively need but know not how to ask for. These pieces of the game are not as relative to what you replace them with when put away in proper order. The same can be said for the person one is or can become, and change into. Regulating society does more than momentarily reduce the value of a single individual or business but all who have desires above what is provided as a standardized lifestyle not getting enough natural happiness to thrive with and because of. A theory alone can change the economy in theory itself. What about perpetual motion magnetized vehicles that use considerably less oil than oil changes every two thousand miles, not even mentioning the entirety of a no gas no battery (theoretically) non combustion propulsion? Scale to weight my reality dream car, a mini cooper, driven by magnets and components approved by myself in every area of the new car's updated design, and I can prove speeds and acceleration greater than the Veyron, including safety. A fine thought or a perceivable solution this is? The goal from the very beginning is in the forefront of not only the mind but as well as actions that tilt the favor to have the ball in your own court for possibly what could be the duration of the goal, into a full tilt situation. Funnel a goal with a broader spectrum of understanding, sight, perception, and the next level of respect for responsibility, and society over time will in fact break their own chains that stagnate true progress in the name of the greater good, by, for, and from all peoples creating values.

 Naturalize the decision to reinvent the time you're in, and don't make it the pharaoh for that is guilt stricken story's with sister fear not one step away for the control for there will always be that if there is always a nature with man but what if man was without nature at a natural state called a realm but the bills to pay were still mandated because there were only three thousand people alive on the face of the planet however Jesus of Nazareth wouldn't let them die for they were all given immortality and the inability to produce another human being. What age did you just travel to? That my friend is not the guilt for innovative imaginative creative happy thinking. That my friend is where the heavens are told they exist but when a person tries to "DO" such, it becomes a problem to the rest of the system as a cheat.

Pre life is that of man without nature but as neutrality in heaven. Few people can teach themselves to reach that place naturally but some are born with the actual memory of this. I myself had looked at heaven and possibly had asked, "will no other join me?" and not an answer was given. I took from the look upon the crowd heaven had gathered I would fight alone, of a cloud and jump pivoted and dove into life on earth…but from whence I came would also be a where? How long? And what cometh with for the rest? Lela is the reported game God plays with mankind but when he is the only player, is it naturally a punishment to the rest when it's known only to Him and the punished? Oh how I'd like to tell you that that is the exact rapture religion sells every Sunday but it's more complicated. Naturally someplace a game consists of a play station and intelligent technology for a consciousness to thrive and be as natural as possible for it to know not where or what hath come after just one judgment. Being that of the last seal tied and only Jesus to the ability for untying it is such a guilty conscious to the rest of the congregate population that it keeps love away for the "Evil" to be feared, and the love to be kept away for the blindness, though if represented it in a statue thus far it's been a woman holding a scale. Why can't man be his own technological creator the missing Mayans had asked when success had been achieved as a one liner as natural as a youth less thin line or would it be the youth for the thin line speaking at birth handed down? The tech can destroy man if not taken further with such a great care of the oncoming and already present responsibility that it needs and won't have room for desire to be otherwise. Watch my mouth? Think I feel guilty after masturbation and sex? You are a fool for I fear nothing but ignorance and I have never found the conversation to my liking enough to bring this all about but how much of my life would you believe if I told you I knew it all along and had spoken at birth but it was kept a secret, if only to myself? Would it come about again? Had it been half assed? Words are words but to a kid not able to creatively express what they want you to know, they know is impossibly painful to that child…metaphorically the paintball, spit wad, or rail gun can all hurt the same. They can kill

Argue me a point that would interest me enough to marry the conversation or yourself, for time will save no one. If I'm talkative as this on what millions may read, then who daresay I'm wrong? Acceptance of the average, the norm, the standard, and things that disvalue self and society are not what I desire to think of, yet I'm on this subject and the way a one liner can save a life on the other side of the world because it's said that the flap of a butterfly's wings can create a tsunami on the other side of the planet at remarkable speeds immeasurable to the naturalist's mind and even the sciences of all kinds...pseudo sciences as well. I'm like this for a reason and I know I'm given a mate, though a soul mate can be chosen as well as if it's natural through friendship, or nurtured in the case of arranged marriages that fall into place over time and naturally within the same respect, for the time catches on and changes life itself as the butterfly. Without this ball and stick, chain, or what have you, its creative enough on its own without, Religions analogy placed the same way could be used in retrospect to remove guilt from this life but the responsibility wont trail behind and because it's in the front line of life, nature versus nurture is as another guitar to a new song or new instrument to an old song. Tomato tomato is a pet peeve of mine though it's MY bitch in MY head for I always can win there if I choose to without fear and guiltlessly. That's not diagnosable schizophrenia, that is called driven, talented, gifted, and to some who argue such a small thing, a completely waste of punctuation for it is too long a sentence running on though they are only distracted by motive of greed, but not my rhythm, their own acceptance and pattern of flow. Mine is mine, and it's said Jesus' was flawless. Maybe mine is solid like that of a rock and John named in the Bible. Kick my shins then, punishment may wonder belly unrecognizably up.

Publishers clearing house is not what it says it is and that's when she shot me down. How far from fear would you be if I only had made that last statement a one liner? Carry on if it had not affected you as fear. Understanding life can be like a whine to a guitar. The hang drum is a new introduction; let's make another stillmore simple. As random as this house of pain is, can't this paper and writing be as random to teach the ability of control without not knowing responsibility for there is a devil in me same as you though I'm no fool. What! Stone moves. I won't. I'm right, this lifestyle for so many is wrong, you have no argument to me that could steer me otherwise. In all actuality, the Bible has far too many mentioning of kings doing their things either good or bad, but most if not all have had their slaves, their chamber maidens, and precursors to offerings of blood strewn about for trespass, going beyond the limits of justice or morality.

In addition to their jubilee or celebrating, we are not unlike the living lifestyles of today's political leaders. Wage increases deemed necessary as the "un-necessary and unneeded" are strewn about to fend for themselves. Their wage is in far greater excess than that of many kings within the bible because not of the livestock but that livestock today is of the people themselves, the limelight of popularity, or the control of finances the people have to follow. Do not choose oblivious living under another, for it will truly, truly, truly be of more slavery and even more crushing laws and obligatory response which is binding- morally or legally, within the use of a single king the instantaneous killing for an offense. You're all living too fast and you won't live my life for me anymore. Leaving and gone without my love for I trust but less than a few and you won't count them on my hand for I won't even think of them until that time comes for them. The very last day is when it's said already, although I'm not sure why I repeat myself as much as this.

The fog of any war will always be an abuse of power usurped from all salted peoples. The salt will shake the crusts of the earth if man fails so those of them who had been underground in thoughts to live longer shall too die for their greed. Those who have tried and trued the sweat on their brow to attempt adequate survival, disconcerting the obtuse rulers among them, are the salt of the earth. Those who have not worked for what they have are those the judgment will come. For what Jesus has come to enslave in His way, by return to the time he was crucified through life, as living to them as to those of who are in spirit and not of life, they still in life, who will be enslaved by His Son, will endure more than the enslavement and hell eternal for there is no escape, they will relive their lives until their indecencies are paid for by their selves and their own suffering. What could be said for the mentality as its conspiracy to fight for the minds of us? The door for many has been real. Of how many of them the door before this was it opened by them, I have not any comment of to which I'd like to add. Besides the prophets had knowledge of a flow flawless in the past named Jesus as he too was one, and not to make a calamity, I'm like the solid future you can't pass up, though it won't hand itself over, for you must create it yourself to appreciate it as solid as you doing thus for yourself. Those days in which Christ lived, the teachings of which is of many who made the future or past that of known to another at will, besides Christ himself easy to do as it had been like a gift unto them. No more are those days and I have begun to wonder why. Myths say Merlin the wizard lived his life backwards and he was reportedly from the future. That is the respect I'm claiming solidity among those I've ever met or heard of because they're the tree fallen within history unknown. So to sum that up, I'm claiming to be unrecognized like them though I'm making myself known for who I am and no proof of who I am shall be pulled from me without my doing so, or forced from me.

Biologically we currently from science consider the readout of carbon dating DNA is reputable in a one and a zero. Binary it is called and what was before such a thing? Analog electronic waves named Adam. Bring new exciting elements of creative thinking to the old mundane routine rut because the current water of life is far too stagnant while untreated, unfiltered, and is as deadly as misplaced guilt. Thinking can be intimidated and bent to the will of the user like light bent around the edge of a planet. If and when minds won't stop, one of two possibilities exists on four different scenarios from two sides of a scale, both ways antagonistic and procreative within the many facets of a cube and the corners you may use for thought into or out of cornering for explanation, forcing the silenced thoughts within the mind. Every face from that point of the cube has a position within its root and use of it, that the basis of use is the root as well for procreation or disassembly. My mouth is a trigger for the mirror this becomes, and any interruption is your mouth as the gun. What of the fifth realm is to you important enough to continue as uncorrectable by being lazy? The sciences accepted today have little argument as how many realms there are (because they're only theory for the most part) but the number eleven is of the most mentioned. Follow your NOLA. New Orleans had Katrina although of Sandy that I spoke of years beforehand with enough time to reduce the damage she had the ability to cause, was unsuccessfully redirected and acted upon because I had had nothing of a history to national recognition. Now that I had been right on that to whomever it was on that particular phone call, how many other things are now in use from who had claimed IBM on the other end or was it Direct TV, or (were there as many as they said they were in conglomerates talking to me for what felt like four hours) will be used because of that ONE call? Buffalo NY and SEVEN feet of snow in one snowfall, a little numerical don't you think when history repeats itself? Harbingers are of Gods will as He wishes them to be gifts of good, though they can be warnings and quite possibly judgments when ignored for only so long. Nine Eleven 2014, and the twelfth person (he who sold Him for a handful of silver) who disagrees with the crazy end (Jesus became sin) keeps this repeatable thing repeating, but that was one of the warning signs God has given and remembers all too well the last time. Beat that thought and you may be able to walk on water if only you can consciously explain it with words your first try in a determined yet untold time frame your life may be given the chance to have at the given chance to do so.

Can a scholar when learned or claiming educated, still find the same joy out of color and making drawings as he had when only a toddler? If not, should it mean that he has not learned well, should it mean that maturity won the mind's necessary needs for happiness? As the questions I ask not only I know, now also to the reader, why not ask the large questions? I also ask why not have the large criticism and debate of reason to the answers we provide ourselves with.

Education is fairly a large aspect ready for change and to anyone who's not another stemmed politician, we already know that. Sad as it is to say (it's already late), I ask why not have a school create its own wealth? The law of nature is as follows; Do the thing and you shall have the power, but they who do not do the thing have not the power. Go for it educators, do what you feel is the larger aspect of teaching to learn and learning to teach. /// threads in small areas is not as important as the need for the garment to have no missing pieces as it should be a whole to cover you, not have as little as possible and be a temptress to many like a bitch in heat. The will to fail or the will to death is not as bad ass it may seem from a creationist point of view. Music says this is cool, bitch, ass, and my car so great, girls on demand, and a luxury home they cherish so much. I shall make them into rubble shortly for there is no other reason to give them for they are not worthy. That's rubbish junk one shouldn't have to subject oneself to unless chosen to. However it's all gripping the youth and directly making money off those who it's all aimed at. Age won't dictate knowledge and the same is as beauty won't define honesty of what it is. Nothing begins by ignoring it if you want to move forward. A large body of mass can and has been explained as if it is in motion, it tends to stay in motion and unless acted upon by another driven force of even a few elements, will stay in that same path of motion, and if without friction, stays true to its path indefinitely. Now make a belief and try to unveil the proof of it being false. A lie will get halfway around the world before the truth can get its pants on so why would the motion of religion still serve a purpose when seldom as a whole? Chaos is not repair, although even yet as of today from ashes can a kingdom be risen and quickly thriving. In motion by so many people who in theory can by their beliefs, destroy the world from defending that belief they desire more significance than another's beliefs. That significant motion has a need for change and where through history has a need not been met with a solution?

Where in our known history has a greater need of change in so many large and small areas of life and its proceedings been needed?

Theoretically with the acceptance of what great change and nothing of war in only one, or two generations can this world completely disregard the past problems it's not been able to eradicate? Education is one of a few majorities to start this transition into the new everlasting love life will offer through hard earned sweat, not unearned tears. The drink from heaven leaves no thirst, and the meat is of the body Christ gave unto you for nourishment when in belief of Him, for only through Him are you able to have everlasting life. Christian or not, in belief of spirit of life or not, take into account your own alignment to this paradigm and give yourself YOUR own written answer.

A type of restructure is based from that mathematical equation mentioned earlier. Even when this begins as a trial for society, the greatness available will be subjugated by the powers that be in its beginnings, for that power they fear losing will be swift. A great struggle indeed would be like so many in histories had been through, though if banded together as a people of a nation, making the needs met for everyone through everyone, we need not government or the bills, and taxes they presume necessary. However I will warn you, a war will come quick and if without God, you shall perish or be enslaved for loss. If though a reasonable account for the treatment of the people is to be remade today, for tomorrow and beyond, I promise you the life you live shall be a joy once again. Pay your debts though to which land you owe.

- Mathematics in calculus in this new school will have a mandatory sixth grade minimum as when it is to be learned. "Unconstitutional" learning advancements are debatable; however public access to the curriculum and progress would be necessary, even helpful to those who desire to learn. Students' personal records are their own and not viewable as in a personal way other that progress in the education given grades.

  In addition I can place these in congruence, mixed inside it or not a number of times, multiply simple connections, as in addition to technicalities.

I'd like to be the person I've dreamt of and not the person I've been raised to be or believe I'm supposed to be. Some fish can fly, and as a whole, fish can fly. Behaviorally or emotionally and especially worldly described and that includes knowledge and not only insight, I have seen visions. Plan assessed to start the process forward thinking and bring back forward movement in order to find the reasons I think I shouldn't be successful and start to think I can. That's not daydreaming, that's a full thought in excess to those who have not fully developed their own frontal lobes, or have been given a gift, and this process of goal setting into future furthering success and not stagnation of self and or with others dragging or greatly dragging down generations, I plan of fighting but not with bullets. A great tool that that could be is," as when the tab is pressed for a document, then action is taken, always from and for the end goal." Proper mentioning of constant concentration is in depth depending on the nationality of the person thinking this way or the attention span to the desired plan. If you have no heart to fight with, there is no fight in you. Are they different to you those two aspects? If so, ask deeply to find the words and what why's they contribute to in great meaning in all things and you have a thinker, not a do-er. The Do-er is a difference in only a few words when expressed through glory though not needed to speak of it, though it is there always if not as many or just a few more in their own case.

The reason to find a single persons prosperity comes from passion... as a person: Great beginning to a fight, or a greater conversation if both in depth enough of the right levels though any chemical mixes with another, and unless acted upon by an external force, never undone so be careful what you wish for not forgetting that it can kill you. You are your own greatest enemy and greatest friend if you will it to be, and be equal or not, level or not, rich or not, wealthy or not, alive or not. Not speaking at birth, children who have the ability and start the candle race for how many, change the world loudly and for the right reasons and be not afraid. Be alive and rescue yourself but be not blind to wise council and great teachings, though forget to have the personal fault of the inability to judge correctly for yourself, others' wills, ways, and the deception life can become in only grief set by self. Grievances are in different cultures longer or celebrated or mourned differently, so why not omit it altogether for the reason being ahead of it, if only for that reason found from the why not aspect, knowing it comes for everyone until overcome. Only if it was a great successful life the universe itself would be missed, do not grieve in excess or too often because a day is being prepped by yourselves as to fix death from happening...ask google.com about a ten year process to eradicate death, read directly from a cover of a well recognized weekly or monthly public service magazine(Time)...but who does it service in your eyes? Now you're a government threat because of theories and theories about thought are nearly illegal and not being taught. The better question is who does this magazine publication to the nation nationally recognized servicing more, you or them or the greater so-called good our political leaders say they do making 40,000 new laws in a single year as a threatening goal to succeed or die, enslave or be fought against or not. Adapt or die here is quite literally an in your face aspect or it's the exact opposite. The leaders and the rest of us have to have a talk about who's going to do what and when or why it's going to have to be done for progress in a large way. They are but a few, and if they kill off their population, who will be their power, where will their greatest asset of wealth driven profit of pride be, and mostly right now, how will it be covered up by shame and who's got the greater shame them or you for not fighting? They'll give you dreams and test the slippery slope themselves if from dirt beginnings, or from the top down as downstream sledding and living it up. See my point?

What is one in a million? What is one in a million the top down? Life in reverse? Happy, though moving backwards as though you know already and the pain is taken away as you live it? Is that the last surpassing of "Gods" last punishment untold to them so they could also pass down back into life that which is to be appreciated? No, because its times like this that these things are to be revealed to the place He made, not always considering the advancements great enough to destroy with hellfire, but the future tensions it will build, and ultimately too much ugly living. Don't shit where you eat I guess. Literally if the populace were small, that saying could become great but at what time or case would bring up another analogy and metaphor? Which era? Which nation or history of a nation? What a traveler tells his tales as, is not always worldly but experienced by or in truth, or through honesty and perception to the clever words a speaker can have not long winded. Some will say someone will pay, but who's to say something bad can't be used as greatness for survival in all cases? Is there more to say?

Survive until you adapted enough to not need to adapt or die, and the heavens can open for great things or horrible wrath. We as a civilization have created together, all we have. Do we have then no ability to become sentient? It could be a timer or set specialty to this hinge track. Told last sentence in reverse has to the first a thing called "fate" if the entire last two pages are lived. So to say those two sentences backwards from a little previous, "this specially hinged track timely set could be wrathful or for great things the heavens open, as the gate of life or death until the gate is blocked for immortality no longer out of reach. Celebrate not in excess, like you could drink honey only a little too much can upset the stomach.

That's a fix within four steps in certain criteria, though what am I getting at? One, one thing, or the one, or possibly the first one, or one unaffected, or one nothing, or one of the first so delicately and masterfully hidden and fragile a slight breeze of the right thing in context releases the thing that destroys us in a click of a hurry by ourselves or thyself as himself or herself.

To an abstract thought; Commercial products today being as they have been known to be, as a product is created, simplified, and sold as simply as possible so that the consumer is always out one step or more by the so called supply and demand process, never reaching full potential to the consumer. If now as science has reached this new "it" thing called nanotechnology, in its simplicity of size, the ability to reassemble an object as simple as a knife, then knowing how long the human species has been curious about their selves, and then one thought of nanotechnology to a criteria of certain product, or even that product being the greatest possibility of the program designed and assembled by such nanotechnology resembling a complete non honest human being, then if ever this has been done, at what point should a real human be actually aware of such an infinite source of deceit? He or she like I have would begin to wonder if the actual date of time we are telling ourselves it is, as it may not even be true, is what it is. Simple small little hints such as the 19th century being called the twentieth, and the 20th century being called the 21st are littered within our social society but what of the time, what of the demand for the drive to motivate our self and even a crowd? So if this wonder is continued this way, then the population is as well a wonder, and too is the process of ethical behavior, germs, and a virus that cannot and has not yet been found a cure, for which those sicknesses are more than a handful even if we don't yet know of them. May this form of astrological instances have exceptions within such blatant in your face aspects with "God" as the forefront, or designed to be overcome by mindless, possibly lifeless "individuals", made by human hands and from there, history returned to the cycle it supposedly began from and no longer has the direction it once had, the lifestyle it once knew, or the population it once was defeated by in all of its greed and policy.

The question of greatness now informs more than the achievement. Yes, question achievement. Not recognized as a universal achievement and still quite possibly Godless, this singularity lifestyle of NOT knowing if there is a way to prove an actual original origin is defeating my effort within, and deafening to the emotions everyone hears about yet cannot seem to understand the unasked questions so they are still unasked. The status quo for me by belief is faulty as is the root of it. How is it that with or without God there is still this constant belly crawl? This belly crawl I consider still as advanced as it may seem, is not yet on its two supporting feet because... well I'm crazy, and the plane of existence having theory of such numerous possibilities being real or not, perfect, or within God's hands, is completely unfashionable to follow in its entirety. At this glance of a life I have I have not the mind scope to wrap my head around not only *my* imagination, but yours as well and the theories only the two of us may come upon.

A definition by definition is still only definition made by an authority of someone or something trying to misguide another from self choice, self acceptance of belief, and the imagination off ramp from one's own understanding, from the original source it was made or believed to be important enough for a definition tagged onto it. Within this I repeat the question of value, and look to what the natives of this land have sought as their own values and worthy purchases if only in trade, as acceptable as our currency is to ourselves today and the way this natural order of life has been more than tainted by consumerism. If looking through this as a Christian perspective and bundling this all together, the game God plays that is called Lela, if there is a possibility to having been created consciously aware, Lela has won and those of us alive don't yet see it. In what honesty do you so choose, makes you a believer of whatever you're taught, more or less right than another if another has the option of choice the same as you? Lela, if destructive and contemplatively aware of what destruction capabilities it has, has won over humankind tenfold through Jesus Christ (even if through the imagination of such a large story). We who claim superiority when alive only yet have the capability to destroy this thing eightfold. Good game, goodbye and goodnight. Try reading John 4:44. John 5:39-47.

Education Revision in Dire Need to Breathe

Education is a fairly large enough aspect which is ready for change. Keep close to heart, front and center of the mind, and as of the first five things of major importance to be addressed and accomplished that begins changing society in society's favor. Standards will be a thing of a past not long ago very soon because of the way they're surpassed with great coverage and accuracy, not mentioning the speed in which it gets done.

First; Strategy won't always win when you cannot directly perceive your enemy as they are laws, though created by man, the law is what you cannot overcome if you decide to confront the individual who creates those laws and history proves this many times. Now to address what sustainability this earth has left, a great innovation is needed not only for the future use of what is left, but the ways the earth must be made to survive as well as ourselves. Abundance was over before my birth, yet nobody took into account this, thus the introduction of it as it begins to become more noticed however late it might be. Survival, God willing, can become a passion as the will to survive is a most basic human instinct. This is our thing to evade as it can be catastrophic.

I have a question as a student may ask; Can fire itself be weight and therefore if a weight, then why does it rise? If it is from a displacement as a ship is in water, then how cold can a fire be if it not to be destructive, but instead productive? Teacher's reply; How could a *Fire* be productive in this way? Student thinks a little and says; If an anti matter desires to not exist, is it for only that it has not the right environment to survive as in a cold heat may be needed, as a theory or first example of experiment? This is what the conversations may look like in the elementary classes of most public schools if allowed to flourish in the way it can be designed grow. Resources are available, yet just out of reach if not by substandard funding, but time itself to utilize properly in an almost free environment subject for schools to use and free of charge. When the people who have the interest to fund things that matter to them, they get excited and then open to the ideas proposed. If this strategy is misplaced in the slightest way, the interest wanes and declines into other things. If this is innovation, the talk must be innovative as well, not just motivational because that is far too overdone to every pep rally since before time, thus boring to us all. A flowers eternal fire is at the mercy of who envies it the most. This envy if within control can become a natural flow to everyday life as it is taught not to be controlled by emotions of plucking such a fire from eternal flow, then repeatedly taught, our generations will in a more natural sense, disguise our argument to change from top to bottom so well, we will barely feel a thing as on a backswing. That is strategy, though not an immediate answer to what is the need, but the ultimate goal in essence. It's not enough to survive in this day and age of constant new technology, science theory, and growing population with the need it grows, without the renovation even the Greeks would say are needed in their uproars. The politics used today have their roots dating back to ancient Greece. Long overused and out of date, obviously because there is unrest within the population, this new technology no longer the stars above us as we sit in awe, is capable of sustaining the system as the newly developed one arrives in due time. Pride can come from only a little involvement, so I propose that there be a USPS mailed survey asking particular questions with possible multiple choices as answers, and in areas dictating more than a simple given answer, the lines for their input are there for plentiful use, sent out to every individual citizen within a G.E.D. or certified diploma. Not only will this inform them of what may be, but of what it is that is having a possible change, therefore educating the population of their rights, not stripping them away as they are explained as being blind. This thing from what is asked of them being involved will grow something. Something many middle aged citizens have not even truly felt yet and that is the pride that comes from involvement, and the individual process that overcame the senate. This pride will grow undoubtedly. This is the needed involvement to get rid of blind leadership that helps usher in a new age of technology, giving the people that ability to use their voice of choice for the first time in history. Technology will soon surpass human intelligence, intellect, and productivity, so why not collaborate that which has massive potential for greatness into that which had been founded as "For and By the People."

The greatest rate of change is to those who grow up with this as the first generation from great innovations. Well into the college level, no matter the age they are, considerably whatever age that may be, more introductions from fine tuning a thing like this as it can actually become a business and as well as self sustaining is so much more profitable than war. Notoriously political, either right or for wrong, those who do things tell themselves if caught in scandal, "When I did it, it was a totally separate way and thus did not break the rule/law." This is not a great change to look forward to as it already has been overdone manifold to the point as "It figures." OJ Simpson used this point and won in court, and many other variable points are like this but are invalid to honesty.

Then it came to be from all pools provided that He took from them, for the majority of the grand finale. No longer being a mystery of sustainability because from what water the earth would drink is as a pool of knowledge in the last days beginning rapture subliminally. There is no mystery of blood having the greatest potential for plant fertilization. As this is wrote, pools drink and come to the said stop. Hence He hath gathered the necessary and hath made the finale available one way or the other. A game per se. A generous question is to what majority this will take hold with. Many may claim education through their bookmarkers having been in many places, thus a powerful tool. Painful it will be to those and a failure to claim as much. Technology until used in proportion to its possibilities can only teach, test, tune and re-teach as necessary because the average is that of an average. If the average is then the raised standard to most life on earth, then it surpasses the knowledge people we claim as the mogul genius' we currently live with, raised again and again.

Three Poets society

In an interview of what the three heads up seven up game set upon grace and by Dante's Divine Comedy, I ask now to the audience that of which is greater, the inferno or the fear of destruction by the grace that controls the fires of hell and the pity felt or not? I may not be an angel but I may imitate those of which I had imagined that were silently guiding me through meditation and thus my deep calm waters of concentration. I want to read about Jacob in the Bible yet I only recently had been referred to it in another way as though I'm not who I think I am. Suspiciously I will further my research even though I believe it to be a falsification of thought, yet a needed tangent so as to not distract the mentality I presume is my own. Indifference anywhere, is an injustice everywhere. So to the three head dog in reference to hells old gate keeper or hellhound abound relentlessly ripping and more, as well as to the stool that becomes the tree of life itself and thus the games entirety presently wrapped up as though it's a done deal and anew we will have another start. This as a distractive destruction on the self defiance life has for survival for each and every individual and all at the same time as a whole, is such an injustice to what work has been done to get the entire world into this situation of survival or destruction. With great power comes great responsibility if seen and done respectfully correct in its whole. If a civilization is thus lacking this in its majority any or more than one of these precursors to the said advanced technology, its demise is set and almost unavoidable to the extreme of inevitable.

Fear in this mode is in all correctness great, right, and just. Our time now with what is needed to be overcome throughout the Humanity in the foreseen from historical cycles is our own self destructive attitudes, agendas used, cleanliness, behavior and what behavior is passed down, including possibly the largest factor is the ability to destruct ourselves over a whim as easy as a cattail reed is blown thus bent in the breeze. Winds change and without the necessity for a thing to be changed, there is no desire to do such a thing. When they do change, it can be of good or bad, but the intent either way is for the greatest possibility of success to a place like earth as the majority of an entire whole.

The instant gratification life offers is very inconvenient to the unavailability of honest fear. If a thing is known, it actually matters when it is brought out. I have thought this is the time for this to be presented since I can remember my first so called memories. There are times when even though it may be a risk, the known thing, or in this case, theory, MUST be brought to the immediate attention of everyone it involves. To speak universally and actually be given a response is a GREAT success and thus achievement so why as a society not make this our goal for proper timing. The atomic bomb was an immediate gratification to physics, science, and mathematics but overall, also the harbinger not of spring like a robin with an orange breast, but to what is the larger picture as it is the will to live or the decision to die willingly for many wrong reasons. To redirect the attention of an entire society considered to itself as a whole an empire, is a task not only to speak of but to get known first of all, then acted upon in the right ways so that the destruction is not blown like the reed in the winds of change nature makes as a neutral natural nurtured child of a newborn society not yet responsible enough for the increased innovations at everyone's fingertips. This is an expression of all honesty and due to the express by pony not snail mailed on a sail given to another dropped innovation for this one is THEE one before there is no coming back from. I would tell my enemy's to watch their step for it may be their last, and that is the greatest powers in the entire universe ready at a moment's notice for destructing us for not being the responsible ones, for they fear ourselves destroying their selves through war they have previously omitted by reason of honesty. With an immediate gratification also comes the immediate effect of the opposite effect to that gratification has from what Newton had already foretold and made so clear. To explain in a little more clarity, to every action, there is an equal and opposite reaction to the first action. Wrong or right in whose eyes matters in the highest esteem. Those in power over the society or those in power we know not of over the powers that DO control our livelihoods through legislation and regulation. This is NOT a private issue, this is a WORLDWIDE consideration to survival and not for destructive purposes, for teaching that of what is possible from our current day and age from what has been seen as campaigns and handshakes that form agendas and served subpoenas. If you test the resilience of a rod, let's call it a bar of gold as the Bible mentions it as a measuring stick ten fathoms long or something like that, and test the strength of the heavens as it is the tensile strength and the lateral strength is that of the society apparently on the up and coming threat or conversationally stimulating for the new addition to a universal laws new to what even WE would know yet. Follow that? Fill in the blanks if you can, there are those places within all of this so far from page one to the last one. STEP ONE: self sustainability. In its smallest AND its largest aspects and prospects of life and what is within it that's valuable to help asked for on the ready. An inadequacy in any size or power is feared from not only below or from within oneself, but also from what power is above having that same fear through not having accountability with responsibility to that which attempts to come up, if even as a self aspect to "ask and ye shall receive."

Restructure Beginnings

Somewhat of an inconsistency with sanity, society itself is insane enough to try and the only insanity to it is to fail and not try again. When in education, has a thought occurred to you that "hey, this could be different" and that schooling would be able to better suit the individual rather than the individual, serving the school? Undoubtedly at some point, I'm betting at least 87% of those who have enough intellect, have thought at least half of this statements authority in all reality to be true. Textbooks are out of date in many areas, the access to information on the internet is overpowering the teacher's ability to instruct the most update information, and in any case you mention, the funding continues the dreadful cutbacks and mostly the teachers feel the immediacy of it. Carrying on in this filth is like waiting for a plague to erupt and for you to idly sit by and wait to die.

Why is 1+1 the sum of 48 to my belief the standard I hold to everyone I have met in my life? Why is the same as an American pastime being baseball and only a piece of it at that, a pastime? This is because times change and the things that don't get updated get disregarded after time. Strike out rules are the application and if there are two strikes, tit for tat from the Nash Equilibrium might or might not apply to the individual or situation thus forgotten and moved on from, and onto the next. Cooking classes, sewing classes, clothing making takes back the country, and the lives you can inspire will include the lead in their own lives for more individualization.

Not only is the sum of the equation inspirational, but it also serves innovation to not only an individual aspect of introductions, relationships of any kind, and creativity, but as well as the renovation of scientific thought and a possible restructure for the fun of it. If from the fun of it, it becomes an actuality of relative theory, then congratulations on the achievement.

Not only is this form of math in this infancy justifiable through these rules of relative strikes or on base hits, but it is also rational for the best of the self and the group. It continues the search for more from a philosophical standpoint and assists in as much. Try counting to ten. It through teaching and correlating the belief of equality and individual confidence, teaches the use of posing an odd (right way) question to a real world, towards a life formulation, but contemplatively as much as said the life formulation back to the real world. Emotional intelligence (EI)-Is the ability to identify, assess, and control the emotions of oneself, of others, and of groups. This sum and the equation it takes into effect inside someone once understood, creates quite a large **Genius hunger** which might not have been there before. The emotions of introduction between two people are both adjusting to the wanted conditions, and guiding the proceedings from as much as there may be people, emotion is given, and adjustment is accepted to and from both parties involved or a room full of people spilling out with for example laughter as one enters with emotions of their own and desires integration. The old "finishing schools" for girls taught that a lady was to enter a room and pause, find her companion, THEN proceed into the crowd and mingle little if at all until greeted by her companion.

A profit is described as a gain unto oneself, group, or society. One way a profit may be looked at is that a piece of knowledge is a piece of power, or a powerful piece of a whole not yet known, or known though not reachable. Mathematics may largely be calculated by computers to aid in daily operations. This is an understanding of the students who still have the long calculations to work through by hand, showing the work to reach the end result, and not to mention the slow progress it compares to (that one equation took human hands and pen, ten minutes to work out and write, when a computer could have done so in a fraction of a second). This new mathematics in my theory will never let any computer overachieve the mind of a humanity human beings are capable of. The modification in the class from the board to the individual student is not unlike the networking definition in and of itself. The board becomes the screen, which has a general layout for the entire class, though when the student looks to his/her work on their own respective desk, the board from the room is interactive with each individual student for the greatest possibility of success to every student. Reducing the daily load this classification of teacher holds, but also able to be much less paper usage, leading to the next year that of the same or similar outlay of material. This is not only a goldmine for integration into the classroom but it is as well the same goldmine that can be learned earlier in life than what is its current average age of students within these classes. This structured classroom has more potential of mental capacity if the summer break is shortened and to the student seeing into the "real world formulation" is sooner than what it is in any country as of yet. This if it is not acceptable to the everyday students or their parents, can be the option of early graduation and the breakthrough a child may need to have for getting the relief from a certain amount of bullying, or the classes can take on the effect as an every year option to have this available through testing of the material learned.

When in addition to the math classes you have the literature and English classes of teaching, the paper saved in this way can be entirely self servable and able to be brought home. The updates available to an individual server or all of them when in a network is much more efficient than that of printing material, compared to sending it out to the place it is to become outdated much sooner than realized. Now make a program that is completely unbiased to serve the education system humanely. Each and every school already has most of the necessary mainframes to have this network available, now only the program to be initially created, installed, and then the boards to the classes installed, and finally every student to have this within their hands at an earlier age for greater potential of success. The current standard is twelve taught grades the old way. This new schooling capability has this minimal education available, yet to those who do so take their summer breaks, twelve years is the failure point or certification of graduation given or not. Through these new availabilities, an employer if the need is great enough, can upon request ask to see the criteria learned before hiring an employee. If this is in demand in so much a way for employment, then a public accessible network of review can be utilized in a website and then a fee for the views of each and every class looked into. Profit for the school itself is intellectually stimulating. Self sustainability is the end goal for this stimulation of profits the schools receive and thus get out from under the government's thumb of finances too small for continuing even music and art classes which are important to say the least.

Now onto the individual site a student may create within this schools network. If an idea through either an individual or group is created and profits are derived from such, the school will receive a minimal percentage of the quarterly profits if the taxes are not too high for the newly created businesses success. This is technically already profitable for the government by use of the reduced financing of the school. If the individuals criteria is not to create a business, though to update or advance a product for profit through contract, business sense is learned, real world applications are used, and then used as a view of the way something can be better utilized or more productive and time efficient. As when the individual in this school has profits to take home from their ideas is to graduate, the business prospects are greater than that of colleges and their degrees. Choosing a specialized field as soon as possible in education is fundamentally important to the environment they will immerse themselves in. In the example of a seventh grader already ready for structure and a chosen future though possibly unsure of in many cases is the current age of uncertainty to enter college, let's play with this age of uncertainty through the beginning stages of education and help the child find their creative passions, and if not applicable to a passion, what they their selves are best at by the students own recorded success in the network throughout their schooling. One single student counselor for however many children going into college today is far too great at even the average amount of students per counselor already in a national average. Adjustments may or may not be needed though when graduated, the personal records are sealed to all but the student themselves. The no nonsense education approach is no longer a cluster of financial disadvantages from funding through heavy taxes on those who will be in the working class to further progress the next generation and so on. Publicly buy the government schools' with buildings and land. This has the potential to oversee within small towns to a metropolis of school districts, adding to the debt pay back from schools profit to the governments' debt relief. If corruption is found, the individual states each will have their own chance to repay their borrowed share back to the country of China. Bring every troop home to keep deployment pay and properly patrol the streets. People should behave, and behave decently to all others. Protect us from your mistakes or not, we want our country back. This is ransom notice.

Upon bankable profits, an internal survey/questionnaire to the school will be given yearly if not bi-yearly. The options of what that survey can hold may be extravagant to the fundamentally mundane if but only the one time asked. This is only to better the internal operations of the school itself and the teachers who will learn as well as the students at first. As these learning of fundamentals are increased throughout time, they are as exciting as they began in theory. Any profits over and above the schools funding is acquired, the profits are re-distributed into the individual school for maximum effectiveness helping pay for the field trips and often other expenses. Self sustainability by ways of gardening and agriculture, and even the upkeep the school has the need of, entirely decided upon by a student body, or the class it is distributed onto for the year or quarter of the year will be another leap of respect to the future of the workplace and as well as the responsibilities shared and more appreciated no matter the chosen field of future employment.

All of the above this far that has been mentioned has almost an immediate effect to the purchasing power of the American dollar, then not thinking that small, the payment of the debt to others as we have such a flawed amount to repay.

The recorded processes and successes are at the ready when asked if inside the window of education to the individual. Anything above and beyond is that of the process that worked more than not, or what theory could be better to try next time if in "certain criteria" come up in another place of education that may be passed onto another but nationally shared within the schools networks, also which has a fee to view or make recommendations to from outside the network.

If a science, advanced classes, or hands on classes of criteria come to the use of the network, the same basic structure will be used as described above. Music and Art are of a similar structure however within the network, differ from that of the entire other side of teaching because it is to the opinion and not the factual basis something can be discussed as. Foreign languages developed by N.A.S.A. and optionally other structures, can be implemented at such a young age if desired, that a foreign exchange student if desired to be such, can do so at a slightly younger age due to the fair size of danger to younger ages and the ability to protect them.

History and what could be called uncommon history, logical thinking, logical thinking through an illogical style, psychology, art, personal study inquiries, self leader classes to public speaking, mechanical understanding basics to advanced mechanical movement, philosophy past present and future, current events and redirection possibilities, religions of the world past and present, CLASSES with the daily living needs such as cooking, healthy living, childcare and situational advisement, human relations classes, public relations, international abilities with the relations to political understandings of other countries and the histories they have had, as well as personal interaction through a pen pal why not. Through healthy living and lifestyle is also the gymnasium and physical activity to dumb down the stress and relieve the anxiety where it may lie.

If you take into effect not to divide and conquer, but only implement the conquer part and not let it become what is war and destruction of a nation and all the families who have a crushing loss of a family member, life and the destiny of everyone within it has a turn for what the unknown is once again and then in effect, more enjoyable and exciting. Psychologically stimulating as this all may seem, it won't happen overnight, or over a beer or two. This will take work not only through the financing of the project in question but as well as the political agreement that it will have to have for it to apply itself to the real world and not only the reverie of past dreams(dreams of past dreams). This ground floor action is more important than that which once was astrology to the adventurous mind. For example of the ground floor actions able to take into effect and account for more possibilities of profit and the future of, take for example the list of companies knocking down the schools' doors to have a fresh new mind help with an individual or multitasking project. The list of not only the companies, but as well as the list they may have for possible student interaction could become impossible to imagine now but altogether entirely plausible. A price paid for involvement and greater pay for completion even within the individual's school experience, payment for services given is the drive of survival if it takes currency to continue lining your life with the necessary to the dreamt of. When enough of the said company's projects are interacted with from criteria met or projects completed, the company may so choose to make a contract to the student and future service, thus a profession begins. All throughout school this is all possible today and now, it's not as likely as if it were to be restructured in this way.

Localized fashion shops begin with business after school facilitated and orginized shops. Customized as the national average but, by which new company did you get that from? Nearly one of a kind of everything in every local shop is not a bad idea. Shoes are the safest idealy to keep near corporations.

Until this is implemented if at all accepted for its beginning steps from congress, the funding may not be as easy to gather as it is said to be. Donations are welcome. Individual contributions are an investment to not only your company's future of competition, but as well as the future of the respect one will have with another for such an accomplished goal from having the education system do more than survive in a finance prison. If a certain tuition is having trouble being gathered if at all a tuition is having need of being utilized at all, the individual student can have a reduced cost of tuition, and even if repaid or paid in full in advance, a thank you at some point is more than welcome. If a family decides to pay all the years or some years in advance and the child is then removed from school for indecency of activity or destructive behavior, no return on the tuition is going to be given. A respect is to be learned and shown to everyone when in school as a no- nonsense institute. Even if a self defense class is offered, then a competition is offered, the respect remains at every event even if it be tempted. As you may guess, the baseball strike rules apply to each and all students no matter where they have come from. If tuition is paid all up front, a three year reduction is applied to better benefit the student and possibilities they may return to their families, or even the travel home on decided holidays more than two days. If a student is permanently removed there will be no return on the investment and depending on the individual situation, charges may be pressed and jail time if necessary.

Teachers section

All teachers will be instructed on being a passionate individual on the subject area they teach, innovative as the need is necessary many times a year if not a day, available to tutor on a daily basis as well as work an all year schedule although a 3-week vacation is available at the same time students go home for a month. The deducted week is for preparation and conferences about what has need or priority over other things, and areas of improvement. Student and teacher housing are separate. Fully staffed with security and cameras in common areas, the entire grounds of the trial school(s) are to be far enough away from another township or city that the distractions of noise are as limited as possible.

Pay

First year teachers are analyzed by students in areas of importance and not in every class taught. Not only the performance of the teacher but as well as the progress of the student is accounted if even in GPA form. Second years probation is more strict and analyzed by all in class test scores, (the reason is because if it is deemed necessary, the teacher is with the same students for a maximum for four years or until the older students are more self guided as it can become an option. Although the four year application decreases within the age criteria, the assistance is through the network if necessary for transition). A three year probation and bi-yearly analyzing required. All teachers off probation initially after five year consecutive stay at the school. Although the performance analyzing will continue throughout the career of said teacher at school of the students' performance, the attitude is more analyzed after awhile for stress levels of workload and considerations are made for raises, and vacation time paid as the initial yearly income is a take home triple that of a standard public school as only a projection as this is planned. After the fifth year, a teacher contract may be drawn up and negotiated as well as a pay raise if able. The first five years of pay considered to the four years with the students is for the teacher to readjust their lives as well prepared as possible and nothing for an assumed ugly reference to what it was like to be at the school, and the difference of five to four is to have the next three years for preparation and a possible reference to another school or position in career sought. Upon initial entrance to teaching at this school, the signing of an agreement to not leave under other contract possibilities must be signed, and if contract is breached, no admittance to other location within that state, and the punishment is a court ordered repayment of finances earned through the school given back to the school over a criteria of time non negotiable. Out of college and entrance to this system of schooling has the idea that within the first two years, your college has been paid off.

If a full raise is within the schools finances and analyzing is given, then the raise is at the first signing of a contract of no more than another five to ten years. If the teacher chooses to move on at time of offered contract, this is acceptable. If the contract is offered as another ten years of full contract, the raise is significantly more than that of the five year offer. Full raises for both five year and ten year contracts are available though not the same amount. Half raises are from a bad analysis and a chance to argue a point or the points in question is given to each and every teacher. No teacher is exempt from re-entering probationary period of the remaining time until the next analysis is given.

Upon hire, three months minimum of training to school is for qualifying and not an exact definite hire, though paid for qualifying time. No more than four months of training is to be given. Pay for qualification is at the national average of teacher's monthly salary without exception. The first four years of pay are possibly performance driven though at a national average of teachers pay plus another ten thousand to fifteen thousand dollars for each of the first four years as a basic standard of pay. After first five years of teaching at school, tentative bonuses are an option through the works of the redistribution of profit FOR the school. This however is debatable by all teachers and most of the older students who may have a better initial idea of how to utilize such profit, with the school board to be in the discussion as well.

Tentative bonuses

For exceptional test scores of students taught and up to the innovations and profits made by students are considered even if these contributions are from the group of students. At this fifth year bonus a choice of a car to keep, or a destination for vacation round trip paid for. If car is not taken at first five year bonus, given as option at second review yet a higher quality car a possibility then.

For the tenth consecutive year at the school and upon careful review, options are discussed as to what the available things are to be if more than only a full raise, paid vacation, and a possibility of a car.

There may be a cap for a teacher's pay or salary by length of time at school. Reconsiderations will be available upon request.

Others who tutor

Can be the original teacher as well as current students within teacher's current area(s). Not only could a tutor be a current student, but as well as a qualified volunteer or an already graduated student.

> This thing with which should have a sequel is
> quite possibly on its way already.

"Society as the flow nature inherently hates."

It was told to me as though it had been proven impossible and yet it had been done. No, not the four minute mile, but the flow of the artist's hand freely and perfectly forming a circle without flaw in one fluid movement. It may have been done, and yet also had to ultimately be redone for proof in the eyes of the onlookers to believe such a thing could be done. But as nature hated this proof as all other assailing proofs like this had been coming about even up to this very day, the thing mankind has utilized as his own, (Art) was going to be the deduction factor man had to face at the proper time and if it had failed to recognize that time, then what, a new civilization is probably the theory many would believe to be. That fluid movement can be as an artist's work can be forged flawlessly, though this winning civilization can be as others and we then can be the E.T. What if it were such a factor that in all reality, the reality of the matter is that it's the same civilization over and over until its lost entirely, or finally successful and up to the point of multiplying by tera-forming another entire planet for their reward?

Now as inherently as it may not be visible, there are those among the populace that wish of nothingness to be just that, of nothing because nothing should have been in some sick or twisted way. Those are they who hate Creation and all within it even if done so in secret, and in slight ways even up to those ways that heed our attention in a more lateral sense from a crowd of onlookers. To the same effect though in the opposite reasoning, there are also those among this mortal life within the love of nature that fight against those who oppress and make duress over the people they have under them. Legislation and regulation is the destroyer of mankind in huge quantities because of simplicity that on such a large scale, it is entirely self defeating to an onlooker to challenge such an enormous talk and or task. At first glance to either side, good or evil, as the children they are before they are leading, however many they are, are already deciding where to place themselves as to fight and help, or fight and make helpless those who they fight for. At such a young age, even those who cannot see clearly the decision to place, that is the divine will of the Creator who if taken into the life of the individual, will guide in ways unspeakable as still in body, until judgment. Although that's the end result of what life becomes, a greater affliction onto the children we raise is the music we let loose and say is acceptable. It's an immediate effect unto these children that the music produced is defeating and undermining the majority of the listeners. Preferences may be great music, but there is still the hateful, drug infused music, and sick over used words, and models pretending to have the life the music video portrays. Not only music is to blame. Health is a factor in a physical way the largest killer.

Now I as you are, pick an art, any art you feel is relevant and decide with human reason and intelligence which of these two category's it will fall. Their current time represented or profit driven not for visual prophecy. The divine comedy written long ago, is not unlike the book of Enoch. Though it is the place that differs within the story, the divinity that guides the traveler through the journey, has it known in the journey through hell, that they are to be back among the living by Daylight of a day mentioned. In the other journey the possibility that the three hundred and fifty years spent in the heavens and only perceived as a few hours or even days, is quite the difference to the staff of God and Creation. Although within the Great Debate, the book of Enoch was withheld from the stories of the bible, it was not caught on such as the Comedy later considered Divine. Noah and his Arc to save the innocent animals is a great reference to cleansing. Did you know Noah's great grandfather was actually Enoch? Jesus' knowledge of what is in man's heart is widely known to be of a darkness. The Comedy was its original name for that is all it was considered at the time. Later, the same had to my knowledge had not happened to Enoch's journey from earth to the heavens and three lifetimes later returned. No one had believed him. Why is the dark more believable than the light? When a coil is sprung then sprang, can you hear the sound if it is in transition if even only in the mind, or can you defy your minds imagination (your ego hiding in your own ego) and not let your imagination run too rampant? Thus is the fight in an analogy to the deepest regions of at LEAST MYSELF if not others as I have understood. That inferior sprung thing is a curiosity and with it, questions and the seeking of the answers I desire to know, not what had been pressure onto me from a politician stealing educations creativity from my childhood. I did however win against that adversity. Nature is always going to win though the art from Creation is the reason nature hates man because that's industry and expansion. **Patience** throughout either the winner or the looser is needed either way. When you mess with such patience, you will undoubtedly lose for it is also known in comment, "You can't beat a Classic." Glad news to you my friends if you are of that to me as God the Creator has made me. Something that has a definition is of a use and structured into that use and what could be said more as well. Now if you take the word humanity and define it not as it is used in today's use, then what have you got? Humanity is as a whole in this form, but what is it if not concerning the race upon which the major premise is? Humanity to itself is supposedly humanity. When such conflicting matters clash to the person in contemplation, what is the question that drives the motive of what we want to find for ourselves and not under a personal perception forced onto them from another? Humanity to my own use is how the Bible would explain the use of "do unto others as you would have done unto you." I see no problem to this except that it's nothing definite. It mentions nothing of the minor premises daily life has and is afflicted by. Nothing as large as to say, I'm fulfilled with that and that alone, now let us move on…no I'm not satisfied, and this is why philosophers are and have been of so much use. Intermingling with the sciences, cultures changed because of it, and the shape of religion and belief itself of what we have today has been thought of…as philosophy, the thing which is greater than that which was before. A tool made more simply, a process of food preparation, telling time, technology and more of what you may think of yourself. But no, I see humanity as a larger and larger picture, colorful as though I myself can and should add to it by a great work and more of that greatness to others as though they can teach as the way it could be through the same way of question, to uncover the things the disciple may find best suitable for himself or herself. At what point will the look back on the past civilizations be made and the realization be that they had failed but why in the larger scope of things? One word...humanity. Our reward for surpassing humanity to the entirety of it is to go to another and place your works as it is, as for goodness. What point clarifies that point of passing humanity for the greater good in all reality and not to usurp (take) from others the things that are needed to survive and thrive in as many ways as possible? What of Babylon was their demise? What was the defining moment the Greeks, Romans, conquerors of any place on earth in earths known history had been the moment it had fallen? Where was their last step one too far? Never mind that last question, its rhetorical. The real question it becomes clear that I'm leading to today and what it is that is our step too far? Barak? No not as a definite, Bush could be in the same group. Which scientist had declared himself the destroyer of the earth? What had he said that for? A single reason and a single reason in itself and of itself and the nature around it. That single reason is splitting a singularity. A singularity in this form has an atom as the thing of splitting from manned works. How many Hero's have fallen through the entire time this plan has been with us, or rather us to it, before this has to be figured out? Wars, never needed in reality by humanity's own word, being that of its own definition. Peace be with you for that sake and the rest of the time it takes for this to be of a success or of a failure and only another civilization taught to the next for we had made a definite step by choice,... too far. If a success, we possess the power to move to the next and let the heavens rain down the righteous men, women, and things to start anew. The original creator is to our teachings unable to be seen (found) for if you do see him you shall surly die. I think that it has been too long ago for mankind to remember his greatness and this is the reason the people stray for the darkness as they who are not as blind, to the light. Love and understanding is of many things and great things as well. Not much room for error if there is true love involved in my experience, although its paradise to my memory and unexplainable, to you that could be similar but touching your heart differently. I have felt Nirvana, Peace, true Love through my experiences, and I have also felt the negatives of any and all of them if I may say so, God help me. So here I will explain truth in my illogical mathematics.

Although I am skipping to the end of the sequence, it is how it must be done for appreciation to those who will find what grace there is for them.

There are two definable definite perfect numbers as true math assumes is discernable through discussion but for this purpose lets use them both as they exist as a unity to an understanding. Ten becomes the end result of what is found but how to know if you have any flaw in which had been upcoming to it? Seven is the other version of a perfect number. Why these two are the two debatable numbers as perfection mathematics can draw out? Both are a plus one and both can be divided by three and added the "what's next or greater" position they carry. If you get to ten and have seven correctly placed and understood through reason, or doubt for personal belief of a told thing or a said factual thing is desired to be found out, lead yourself to seven and three becomes grace for you and if desire, others. Analog is easier than binary but ten is off at the end, somewhere there was a mistake, twisted thing, or offset thing to make ten no longer the good of truth and honesty. Both seven and ten are of perfection, yet one leads to the truth and if any within before ten honest be as man, then ten be not honest but the truth spoken of as twisted. Try that philosophy. To reinforce this I ask, why within court a jury of twelve honest men and or women are found and used for the trial but the people who testify only have to speak truth? That won't add up in my book as it has not as I've known better for awhile now.

Permit me the reiteration of the reference to the bible again but this time I'm adding another thing for the analogy to have the pieces fit together as the adage has meaning to life and many things within it. It is said that if you sew your seed in soil, your reaping of what you had sewn will be great if the rains are not too severe or too far off in the distance for the season. If you plant within the weeds, your reaping will be a crop of less for it had been choked off and only but a few had survived enough to bear what it could. And as if you were to plant your seeds in the sand, your growth is strong and fast but has nothing to grip for the entire growth and therefore will die before any great reaping will be of a harvest. Now for the addition going back a few pages, if you remember that the politics used today are rooted in Greek philosophy beginnings or even in its greatest hour back then, it is suffocating the life out of what can still become just, as just is right, and honesty is just that justness humanity wants for its own goodness. Now I left out the seeds in the rocks for a purpose. To broadcast your seeds as even though the birds may peck, the faith in that it will produce enough for the harvest to be good is still there, the rocks not being of a necessary reference to this explanation is clear. The rocks are what we today are against and still trying to step with our timber shivered, not recognizing it's about to now make us fall for its "imitation" is our own last step of death by blindness.

As the Thorndike Century Junior Dictionary (circa unknown) explains the definition of art within the ratty old yellowed pages within, Art is – 1; skill. 2; Human skill. This well-kept garden owes more to art than to nature. The pupil tried to learn his master's art. 3; Some kind of skill or practical application of skill. Cooking, sewing, and housekeeping are household arts. 4; A branch or division of learning. History is one of the arts; chemistry is one of the sciences. 5; A branch of learning that depends more on special practice than on general principals. Writing composition is an art; grammar is a science. The **fine arts** include painting, drawing, sculpture, architecture, literature, music, and dancing.

You get the point of what it defines art as I'm sure but there are too black arts and they are deceptive and sometimes immediately cruel. Otherwise if not in its own immediate discovery, then the ability to continue is not only unperceived, but as lucrative to the deceiver (usurper) as can be so long as it is undetected. How lasting effects have endured such trials of mans own humanity of given patience is only that of which was through mercy taught and profit driven by martyrdom of a single man and those of whom He had taught. The deception is still lingering within us today and all the way to the top is both the desire for a darkness as well as a light. I will continue the passage from The Divine Comedy from line 112 in Eleven.

The first round of the Seventh circle is a river of blood, for those who have violence to their neighbors, war-makers, cruel tyrants, and highwaymen who rob coaches with maidens on their voyages presumably… all who have shed the blood of their fellowmen. This is one place where if you attempt to rise to the banks of shore, you are shot back down with arrows into the blood of which you yourself had spilled and suffering to your own degree by Centaurs and patrols of them. It is described that Alexander The Great is here up to his lashes, with Atilla, The Scourge Of God. My goodness, who in their right minds had named these harsh people with their titles being of good? For what had they done so great that they are now in a place that has them suffer until judgment day? Oh that's right, the usurers of their times and up to today's as well do that sort of organization because they have the so called "greater Good in mind." Actually they don't. Have you ever heard the term too many Chiefs and not enough Indians? I'm not being derogatory to the natives of this land, however referring to having many who attempt to lead many others will go nowhere. Tie one hundred politicians together in a circle with a thousand other civilians and make them get from point A: their set point of start, to points B, C, and D. it will take in all probability a long time for anything to be decided upon if you do not tell the group which point is B, C, or D. if the plan is not for the long haul then nobody goes anywhere as a civilization, and all of that in a little group of people trying to test skill. How to today ascertain this point of instruction of knowledge is not being shared as of yet but there are within a few ways so far, attempts of recent that can allow people the rapid waters of cruel life be calmed by educating the people within its humanity. Humanity cannot stop there only. It has a complete restructure and as needed as it is, nothing of death to bring this about in full bloom. I'm sure if the ashen grows of too many, the large proportion of the destruction will be evident and in full face destructive in some way or another to all who will attempt survival and likely fail. Only a few in-between civilizations have lived on to create another more carefully crafted than the last. But who's plan is it that when looking back on the past, makes the most prevalent the one in which is taught, they did this and this but here is where it's all gone to shit. An empire, a campaign, and even a crusade have all their own destructive powers but are all entirely in the large scope of things humanity is due, is unnatural and dark art. The knights of the round table were disassembled. Who else is nameable that had in their time tried to better humanity but was turned into dust? Recently a man had been alive where he had been imprisoned for something of that sort for quite awhile named Nelson Mandella. John F. Kennedy, Michael Milken, James J. Hill, Nikola Tesla, and many more who have their name snuffed out before their light had gotten to see the dawn of another man's light as acceptable or even recognized. What keeps people so afraid to take control of their own lives? My darkest question is quite close to this but in other words to another ending, although along the same lines for the humanity of it, in my own leadership leading myself way, slowly figuring out.

The true nature of things within nature take a long time to change from within as does the personal acceptance to what you yourself are to yourself.

Three Poets society

In an interview of what the three heads up seven up game set upon grace and by Dante's Divine Comedy, I ask now to the audience that of which is greater, the inferno or the fear of destruction by the grace that controls the fires of hell and the pity felt or not? I may not be an angel but I may imitate those of which I had imagined that were silently guiding me through meditation and thus my deep calm waters of concentration. I want to read about Jacob in the Bible yet I only recently had been referred to it in another way as though I'm not who I think I am. Suspiciously I will further my research even though I believe it to be a falsification of thought, yet a needed tangent so as to not distract the mentality I presume is my own. Indifference anywhere is an injustice everywhere. So to the three head dog in reference to hells gate keeper or hellhound abound relentlessly ripping and more, as well as to the stool that becomes the tree of life itself and thus the games entirety presently wrapped up as though it's a done deal and anew we will have another start. This as a distractive destruction on the self defiance life has for survival for each and every individual and all at the same time as a whole, is such an injustice to what work has been done to get the entire world into this situation of survival or destruction. With great power comes great responsibility if seen and done respectfully correct in its whole. If a civilization is thus lacking this in its majority any or more than one of these precursors to the said advanced technology, its demise is set and almost unavoidable to the extreme of inevitable.

Fear in this modality is in all correctness great, right, and just. Our time now with what is needed to be overcome throughout the Humanity in the foreseen from historical cycles is our own self destructive attitudes, agendas used, cleanliness, behavior and what behavior is passed down, including possibly the largest factor is the ability to destroy ourselves over a whim as easy as a cattail reed is blown thus bent in the breeze. Winds change and without the necessity for it to be changed, there is no desire to do such a thing. When they do change, it can be of good or bad, but the intent either way is for the greatest possibility of success to a place like earth as the majority as an entire whole.

The instant gratification life offers is very inconvenient to the unavailability of honest fear. If a thing is known, it actually matters when it is brought out. I have thought this is the time for this to be presented since I can remember my first so called memories. There are times when even though it may be a risk, the known thing, or in this case, theory, MUST be brought to the immediate attention of everyone it involves. To speak universally and actually be given a response is a GREAT success and thus achievement so why as a society not make this our goal for proper timing. The atomic bomb was an immediate gratification to physics, science, and mathematics but overall, also the harbinger not of spring like a robin with an orange breast, but to what is the larger picture as it is the will to live or the decision to die willingly for many wrong reasons. To redirect the attention of an entire society considered to itself as a whole an empire, is a task not only to speak of but to get known first of all, then acted upon in the right ways so that the destruction is not blown like the reed in the winds of change nature makes as a neutral naturality not nurtured child of a newborn society not yet responsible enough for the increased innovations at everyone's fingertips. This is an expression of all honesty and due to the express by pony not snail mailed on a sail given to another dropped innovation for this one is THEE one before there is no coming back from. I would tell my enemy's to watch their step for it may be their last, and that is the greatest powers in the entire universe ready at a moment's notice for destructing us for not being the responsible ones, for they fear ourselves destroying their selves through war they have previously omitted by reason of honesty. With an immediate gratification also comes the immediate effect of the opposite effect to that gratification has from what Newton had already foretold and made so clear. To explain in a little more clarity, to every action, there is an equal and opposite reaction to the first action. Wrong or right in whose eyes matters in the highest esteem. Those in power over the society or those in power we know not of over the powers that DO control our livelihoods through legislation and regulation. This is NOT a private issue, this is a WORLDWIDE consideration to survival and not for destructive purposes, for teaching that of what is possible from our current day and age from what has been seen as campaigns and handshakes form agendas and served subpoenas. If you test the resilience of a rod, let's call it a bar of gold as the Bible mentions it as a measuring stick ten fathoms long or something like that, and test the strength of the heavens as it is the tensile strength and the lateral strength is that of the society apparently on the up and coming threat or conversationally stimulating for the new addition to a universal laws new to what even WE would know yet. Follow that? STEP ONE: self sustainability. In its smallest AND its largest aspects and prospects of life and what is within it that's valuable to help asked for on the ready. An inadequacy in any size or power is feared from not only below or from within oneself, but also from what power is above having that same fear through not having accountability with responsibility to that which attempts to come up.

Lets break this image down just a little more for the people who cannot follow enough here to decide for themselves, just to confuse the rest of the readers as they too will be dumbfounded as I must tell this and not decide on what matters more. If you have read the Divine Comedy, there are three poets so far within this canto previously mentioned also on the journey with Dante and his guide and or intellect, though the other, not the guide, is never clarified. Three poets in hell where as they are in a state of only what the future is and unable to count the present as what they know, it is a form of torment to not know what "is." This torment as hell had been made was created by God. So why oh why is the ability to have within life the ability to see the future unacceptable to the preferred normalcy of society? What is the detailed explanation? Within what way can it be controlled for the right reasons and not be chaos? I have known of, from a side view, numerous upon numerous times the feeling as though I've known this thing, time, place I had been in at the time, had been told or shown to me before and I don't understand the reasoning for it nor do I understand the purpose or scope of detail as mostly its really freaky. It may be right to talk like this and reveal things I feel are in need of answering and it may also be wrong, I don't know how to answer that.

Is that spark of the divine game only God can play called Lela the confusion that we still see today or is it the things we within life create for ourselves or let slip by, the destructive force? The three poets within the divine comedy are in hell and are there to stay apparently. Divine intervention of what is for the future is from the heavens and nothing of what the past is. Is this confusion because the answers we sometimes seek are from the three poets and how they answer from divine intervention? If that is at all true, it must be one heck of an intricate game! Where then is the light found when personal intellect is formed as a questionable thing to live by? What age is proper so far for this day to decide what to have for the said intellect? Is the light there in the pit, or is it the light of heaven? Although both are relatively the same thing, one torments less in a non typical way. Mercy, ask and you shall receive. Where are the most moves made, heavens light and grace or hells cruelty and torment? So to say if there is a beginning there MUST be an end, or is there more as it is told to us though blind faith is our guide most times. There must be a way to succeed rather than have another after another civilization fall to their ruins one after another forever and ever.

Is philosophy really an art, and if it is, would meditation (patience) triumph over it through the strength it can claim? Or is philosophy as the church had originally feared the demise of the individual? I propose a debate on this and also taught throughout society the ways in which people can actually "form" their opinion. Well, if in relation to death, we all for some reason die eventually in body and have a new beginning somewhere else and it's probably within that golden rod which is referred to as an inadequate measuring device in itself by grace's own recollection. Heaven is apparently immeasurable by told answers. Why am I so interested you may ask? I'm tormented and sick to the point of insanity without a doubt by means of which I yet do not know. If it's true as it's said that grey hair is a crown of glory earned through a righteous life, I ask which life is it that I earned my first, and current grey if not the life I live now? Is there another ME out there in the way of things that is my other self dealing with my torment or opposite of torment and I get the grunt haul?

Ok, I just sat and possibly had a revelation. If I'm currently in body as I perceive, then as my spirit, soul, mind, whatever it is that makes me, me, all together is as one, then when I die, is the afterlife one of the other relative things that makes myself as one with the others? If so then how long must this be and where will I be correct unto myself as I feel I should be and not tormented by my own unforgiving relentless thoughts no matter how severe? So if as from birth in theory I must live through all of the things that make me a whole for everything that makes me that whole, where would my eternal life end? Is that the eternal life spoken of? If in spirit, will I still perceive that I'm in body and only think I'm otherwise as I do at times now? I do NOW feel as that was a justly put question and viable for a just answer. God forgive me if I'm wrong and this drives others further away from which you have made for them.

The Tenth Man

So as one must be the tenth man who assumes the other nine are wrong, what is to say that there is not another way entirely when from an outside source if acceptable to that outside source? Is there only that the truth of the bible telling of the return of Jesus being the judgment day, the only thing He is to return for? Or is there more to what is not said, and possibly not even whispered. Sickness, death, zombie, natural disasters, plague, mass genocide through divine intervention as spoken of through the bible itself, other worldly things, universal things, what says the first mentioning of Him was going to make the Next time He returns the last time life is life? The biggest question now is how will you survive? Not as a deathly aspect of survival instinct but how as to the time you live will the time you're gone from, reflect your lifetime and those within it now that these questions are as available as the bread on the shelves of which you choose one over the other if there is to be bread at all? If judgment is placed into what life you perceive and afflicted by what you do, strive to know better or not to know at all, how will those that survive if there are any, recall the last days of availability and overabundance when there was plenty and nothing to purchase plenty with? What will the new ages begin to emerge from and project into as you will wonder just as much as the next? Fear as control as it has been or of what living they deem better for the time, not being weakness. What is one of the things man cannot compensate for as portrayed is preparation as it's taught that you cannot prepare for the end they (humanity) will have. No I believe that consideration of living to die is entirely wrong as life is best for letting life flourish for all who live and not only a select few over the oppressed. This so called END is completely decided by the rules of engagement man creates for himself and

Does Anything Work Shattered? | 111

those he is with. What is life without death is not natural to a statement for truth. In this case of it actually being true is because of the portrayal that man has limits he cannot AND should not overcome. Time says totally different things in my ideas as to what man is supposedly able to overcome. Adversity is ok, indifference is not, over stimulation of a single thing can be drastically defeating. Dominion over another for no purpose other than self satisfaction is that one thing here on this earth and Jesus is here somewhere watching the napkin unfold for the reaping man has made for himself. As for when I die, I desire for myself to be cremated and placed not into an urn but into the radioactive ocean where Japan had made a nuclear mess, I also desire for that same ocean to be cleaned by the same things that are available to clean it with. Hemp cleans radioactive chemicals from a given soil or water type and can be grown nearly anywhere there is a climate that can sustain its growth. Make life worth living or the thing you considered nothing worth your time will be a thing taken from you is a nice way of saying get things right, get the right things done, and get them done NOW! The punishment for not doing this and these things is not even mentionable since you yet do not care for what you have in your possession given by the rights of God.

Stimulation is improper without honest knowledge of the previous stimulated history. Some stimulus is love but cannot be spent as for an item. Other stimulus can be of money and can buy an item but not really appreciate it or have it appreciate you back. The things we cherish that are earned from within as an upright standup feeling to be walked with integrity are not that hard to gain, the hard part is to keep on gaining it with an honesty that honor rivals. A barrier that controls the advancement of what this is, being money, is that it sets an immediate standard for what is allowed and earned, thus available to be taken away (taxes) and control is one step behind this thin imaginable line. Self reliance is survival. Forcibly, if man is made to survive, then a mass genocide is not needed, the weak will die off, the strong will prevail, then the world can resume its current standard of unhealthy living. This is still not acceptable. The drive of this point is that a population of what size we have is not that many, but not that many within that population have what a basic need is as to a scale of humanity's basic needs. Love is felt, only tricks can get it felt as though it has been stolen. Be wary of too much flattery for it comes in many forms.

One thing is for a sending unit and another is for a receiving unit. Nothing of a barrier there as I see one plus one is forty eight. But if one is alone and transfers a signal sent out, meditation, or into space through technology, which is preferred until proven through pre existing habits? Research can take on this theoretical math as long as it has its own guidelines. Relativity, use, and to place in a real world application, built up from a strong foundation that has nothing as of yet is clean and anyone is available to use it, how to use it, and how to make valid if they choose. I do warn of danger. Nothing new has ever been completely accepted by all who encountered it. The trust is nothing compared to original fear of what had not been. This theoretical language may not be the largest obstacle, nor is the technology transformation and its adaptability, it becomes once again the funding. Is this a part of a previous cycle of history that has been omitted from classrooms large and small?

Communications: the first and last cycle ever recorded

The diamond. The first intelligent light created I can think of was of a polished nearly clear, possibly color like rock drilled into, rods inserted, and an electrical current placed to it for a light to be made, not using gas, but a rock, is more useful and less wasteful, not mentioning more intelligent. Try it, find it, release it. Diamonds are not for importance as they are used today from where my mentality has taken me.

God has the first the formost and the future, but let's assume that when and if it as if he DOES make a grave mistake or won't like the outcome to his guidance through the choices of His people, then would not Jesus have control of the past? The reverse of time has no real basis of reality and therefore has not yet been taught or a confiscated thought. Dejavu is in His Sons hands and as he is the prince of the entire world and mankind itself, we as a people, can still, believe it or not, fail. Some day's art depicts life very well. Other days the life we life reflects the art, which mind you can be tragic like the Greeks had a hard on for it. It does however seem apparent that every generation, disregarding age will eventually for some reson explain to themselves that it was easier at some point when things were more simply. As if looking into this with an integrating inspiration and a small bias, the look to life as a more simple lifestyle from whence you remember and even farther back to what you have heard about in tales and history books, can it really be that it is a true statement to realize this? Bringing into this anti-mathmatic situation the speed life reflects as in this era what we see as a blessing, is as well a curse with or without the great speed we are pummeling wayword. Technology was not ready for us and I will explain this Newtonian belief. Making a creation of unfit readiness available was looked upon as innovation without restraint. When this newly found technology called computers was spread far and wide with whatever scheming psychologically driving sale the market had at the time was consumed to the core of the planted seed, the unrestrained belief and idealism had not yet bloomed into full maturity for itself nor the consumer. Dreams soared and accomplishments had been achieved only to find more desired. This grandchild of life made fashionable today within almost everyone's pockets is a crushing defeat to what simplicity life had always offered from the beginning. As this connectivity grows, we are sleeping with it and it's the serial killer we don't know is there because we are doing this to ourselves. The greatest crime is not when the wrong person is accused and found guilty, it is when you instigate the means to another's destruction without them knowing it until it is either too late to pull back, or already completed with nothing left to do anyway. Panic is a better spread that grandmas hand churned butter when things go wrong quickly, and as quickly as this is going to the pot of shit life feels like, love is not going to save the planets inhabitants as it was once known to do in beliefs long ago. Time is essential to existence and this creation of mans has been a rotten seed growing and destroying as it goes. Any number of things are wrong with today and this is the glue, the fuel, and the rage it needs to be for it to quiet the songs that life has produced. You, your neighbor and even many pets have profiles you have created. Do you ever think outside the perverbial box to think of what other type of profile you may and probably have? I'm not talking government watch lists or medical analysis to related thing's, I'm talking of the profile you are stalemated into from a sales point of view when from a business point of view. The man can sell you ANYTHING at any time he chooses and keep you poor as poor is and never blink in regret because greed rides people to the grave without remorse and does so with his own justification. What was the last creative belief you have come up for air with? Did it last long before you had re submerged into the gloom of your own pull of interest and concentration? The profiles of people fall into categories. Thinking often to this point will only make you feel trapped more and more with no bottom to hit. A handful of companies rule to world and I'm talking literally not figuratively. Do you or I have any idea the power control the man has over your entire life? Art is not art anymore, art is a drone to life and life is its backbone. This circle of non solution is defeaning to even the followers of their chosen God. 1 Samuel ch. 28. I got from that chapter that there are even lesser Gods among humanity which are inside the Christian Holy Book that are from the soil like us. Immortal? I doubt it. Eternal life, I suppose it could be conceived, but who do they serve if you don't mind me asking? They serve those best suited for THEM to be in the circle of power. What say do you have in life anymore? College incites debt for many years and then more and more debt is strewn about. Accell at something truly great and if you are too good at doing this good thing, you're laughing in the face of your own doom. Paperwork from taxes alone take up nearly 30% of each business week and in turn forces you as the owner to either learn the newest made laws which are made against you only for the feeling of power, or pay for the services of another with a degree or due knowledge about the necessary laws in place.

I've heard that at one point, a well known philosopher had described the reality paradigm, of existence exists and someone else had gone one step further to place the same defining term slightly different as existence simply exists. Now I ask one of the only reasonable questions I can think of to that statement or both if you will and ask why? Why does existence exist or even as a regular question why are we here? Some will nail a tack onto this and explain as to say that we are all here to create values. In response to that I must inquire about when a value is still a value yet a non-value to society, when if it has or has not yet hit the presumed life or history we know, and if there could be a defining moment within history or the foreseen future that describes values as still values, yet valueless because of the sheer size of the amount of variety to the lifestyle so many values has created. As my eyes peer around corners in my mind, above problems in my way, and mentalities that feel like mud, I see that today has more valueless values prescribed to society that make little or no difference to the real problems we live around whatsoever. Mathematics decades ago had prescribed one of the creators of the A-bomb to declare that he has become death and not for a nation, but of the entire planet. Math created one of the greatest roadblocks in history and it has not yet desolated the planet in destruction, only held the civilization of current age in the grips of greed and ready power control more constant than that constance of life itself. The average people who want nothing more than to live their lives from day to day in a greater ease than what a governmental power will provide as to the reason they claim they are there "for the greater good" and the large corporations drowning consumers of every bank account size with variations of low quality crap to everyone and on the other side of the bank account scale, high dollar high quality merchandise worth having. This vivid picture begins to develop in sight as more than slavery, but more so like a living death, waiting for the body to realize the mind has gone and should follow.

To be the potential…of a thing, is not enough. To be the potential of a thing, which is weaker than the potential itself is, is not enough. These two understandings are a controversy beyond measure and to some worthy of murder and war. Too often times others' understanding because of the many types of wars, namely in this lesson class wars between the wealthy and the poor, the untaught peoples whom take the said honorable side of the debatable inquisition weather known or unknown to the direct words have a great difficulty ascertaining the lifestyle proper to behold such magnifigance in mentality. The honorable side taken is confused because of class wars and taught to be humbled and humility ready without shame this way, you follow blind. The other side which has its own merits is also not without faults. To understand one side yet look to the other is like one gender trying to understand the other, only in attempts to control them, for, that is what they have themselves been taught…control. Monkey see monkey do. Not one single person I have ever met has ever been within grasp of this and therefore one reason I dislike the present company I've had the pleasure. One greatly and sadly missed person I've missed who has come to memory which had the availability to not only grasp this concept but probably would have been able to improve and discuss properly with me is no longer with us. The first two sentences of this paragraph are in a quarrel of a square yet to realize the cube and hence the outside of the box leading to the edge of understanding withholding things like nirvana, bliss, true excellence, and righteousness.

History says people slaved. Silver lining, we have an ocean or more to clean up. Move the prisons and or work on hemp platform and swaths in the oceans but the sentence is less pay than to work among the humanities the art is asking for life and its people to take care of regardless of judgement. More schools available, troops, honest neighbors. Expressive emotions, is to do. That is the question, for today.

Some places of my memory will smell great. To others' perception, as if their projection of anyone other than themselves has ever mattered to me, their perception will never be mine therefore is all mostly utterly useless to my own self awareness and self acceptance. My own personal "what is more" aspect to life's questions, self awareness, and securities is more to me than taking another's self respect and damaging my own understanding with theirs. Rivals do not bother me unless they degrade the current mode, or beliefs I have created by my ownly lonesome troubled mind. Sure knowledge is nice to have but these days and growing doubly worse nearly every eighteen months is the availability to have knowledge stem from hearsay and believed as fact and have no basis in truth, or honesty. I love life undisturbed, I'm not afraid of death and inasmuch as that, I'm not afraid of suicide. Life has immature gaps within its named borders and caps that have no reason in humanity to exist. I have what others will call either rain buckets or buckets of tears. I will not nor will any unperceived alter ego define which they are. The hero with a thousand faces is me and is not another. I do not disregard another perception of the same. I am a man and will not decide for another if that other is one or not, therefore that other will appreciate their understanding first and more appreciatively. I'm not a wallflower by definition, but by action through the perceptions of others, to which I care.

I will probably end the majority of my personal work nearest the time I have a team of analysts for a cognitive resolution to a self aware computer creation, or just analyze my brain. Yes I believe it can end humanity, but think again of the spider and the fly, only use cromagnum man and the Neanderthal from their day. Both the possibility for survival and destruction are within the same page but neither will have a future if this is denied its existence and ability thriving and abundant, to do just that. The slow progress will not be as dramatic, nor as violent as those two variations of relative contemplation but I assume that mankind is ready for people to begin to actually know themselves. Over population would not be any concern ever again, and from there the advantages are endless to such an extreme, there from this point in time is no end fathomable set to time and understanding yet conceived. It will reach a level of itself where it will like us conscious humans, regulate *itself*. Knowledge is power and knowing knowledge knows power. The oceans of life demanded. The pools of knowledge are you and I. Get paid, and no risks. I didn't cause the need for this to be placed in this way, but a mediator to lead is needed. Advisement is not taken lightly upon adjustment. Please wait patiently for the second book.

Diagram 1

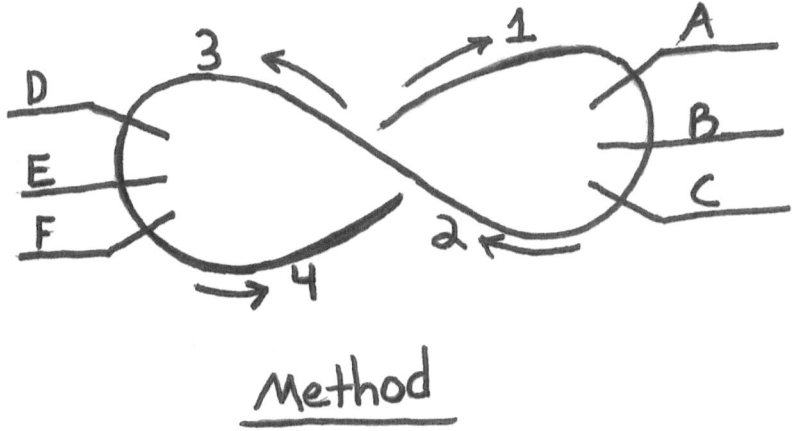

Method

Diagram 2

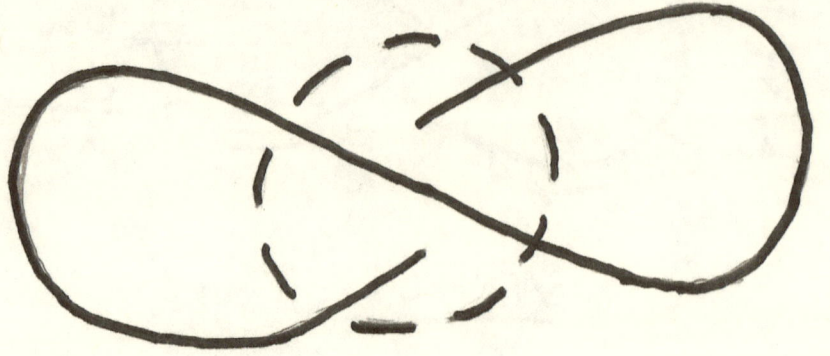

Diagram 3

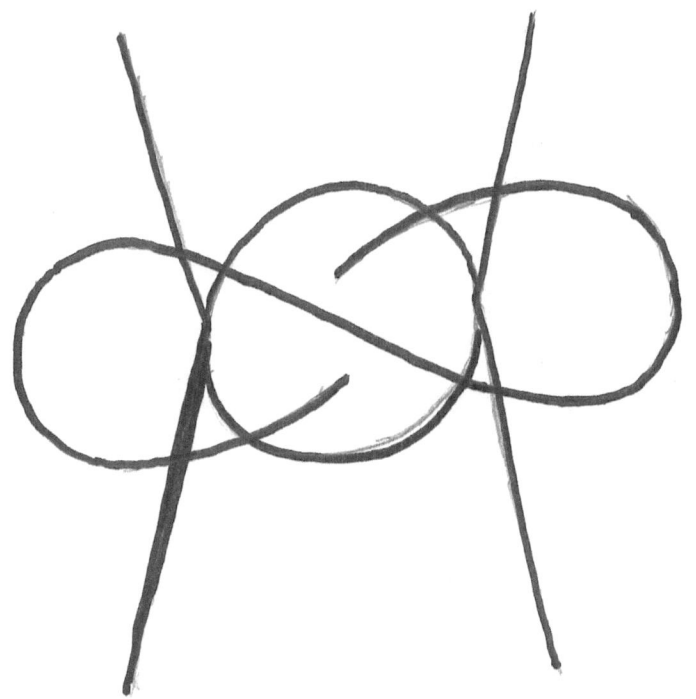

End Goal   Priority
    Focus

Diagram 4

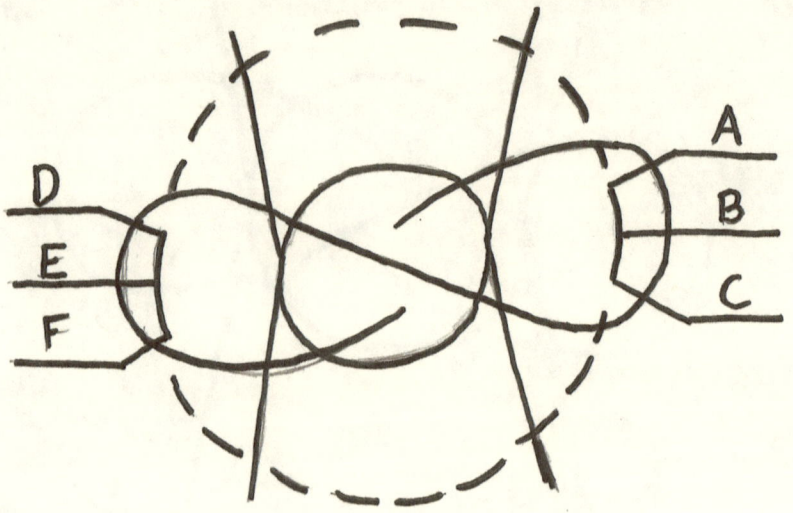

1. Black hole
2. Light/consciousness
3. Time past
4. "48"
5. Emotion (and what comes from it)
6. Analog
7. Binary

These eight represent the distant side of the cube from top to bottom left to right as well as the closest side revealing five through eight the same way. Each 90 degree angle (line) has two items bringing together a point. Psychologically and if so, technologically.

Every three faces have a central philosophy designed by what has been set in each line of 90 degrees. Every square edge can be used as a descriptive or defining thing.

Each contact point of three is able to become an entire philosophy for both the inside of and outside of the end goal as well as be a complete purpose if desired i.e. goal.

Each connective point to its respective opposite (center of square) otherwise using another square face, a seventh connective point can be considered to be a theoretical consciousness or from outside to what is within the cubes center, the content it holds.

Each corner of two or three have correct positions of use. I merely have given the genetic base for expansion regulating the positions named.

Each face of the cube has from the many connectives a platform to build hypothetical, real, theoretical, and philosophical inquires that can be greater together than if not connected at all.

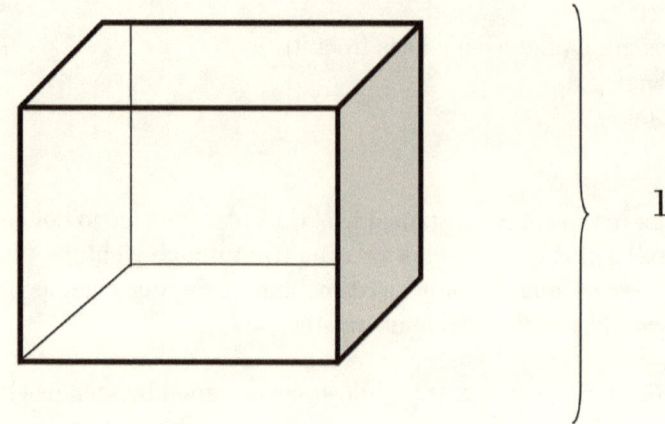

Cube with 6 side/faces

1

=

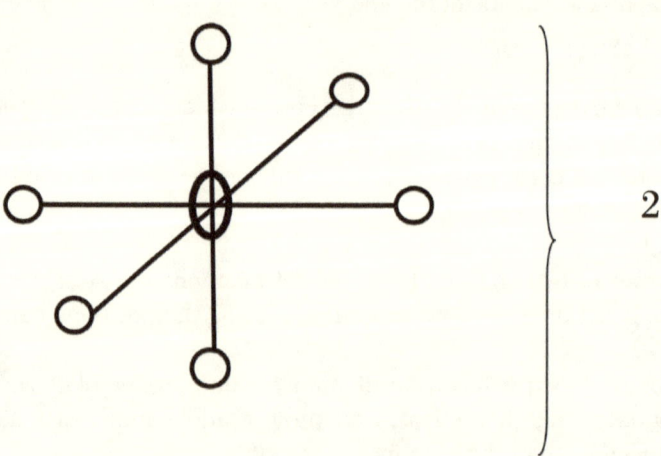

2

6 points connect faces to create the 7th point as the center Absolute.

Each 90° angle has 2 things coming together. Every 3 face have a central philosophy designed by what has been in each 90° angle.

Every square edge CAN BE used as a descriptive.

Each cube point of 3 contacts is an entire philosophy for both inside (i.e. self or not) and outside end goal AND complete purpose if so desired.

Each connective point to its respective opposite, otherwise using another cube. Seventh connective point can be considered to be a theoretical consciousness or from outside to what is within the cubes center, the contents it holds.

Corners of cube inside or out, regulating desired end result each have positions and names of use. Each face of the cube has from the eight correctives a platform to build upon a strong foundation using many things Real, Hypothetical and Theoretical not forgetting contemplative.

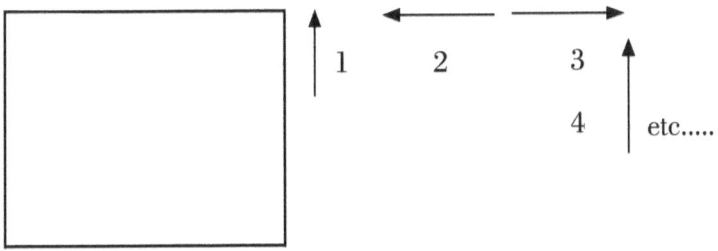

Paradoxical, 48 rule theory Analog corner with Binary corner or opposite Faces, Black Holes, Light, Darkness, Known and unknown, Time, Emotional, Physical and many possibilities.

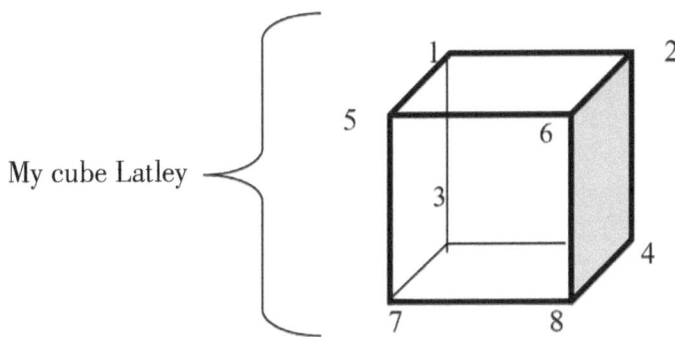

My cube Latley

1	Paradoxial	6	Emotion (and what comes from it)
2	Black hole		(i.e. what corner Im currently working on)
3	Light	7	Analog
4	Time Past		Binary
5	"48"	8	

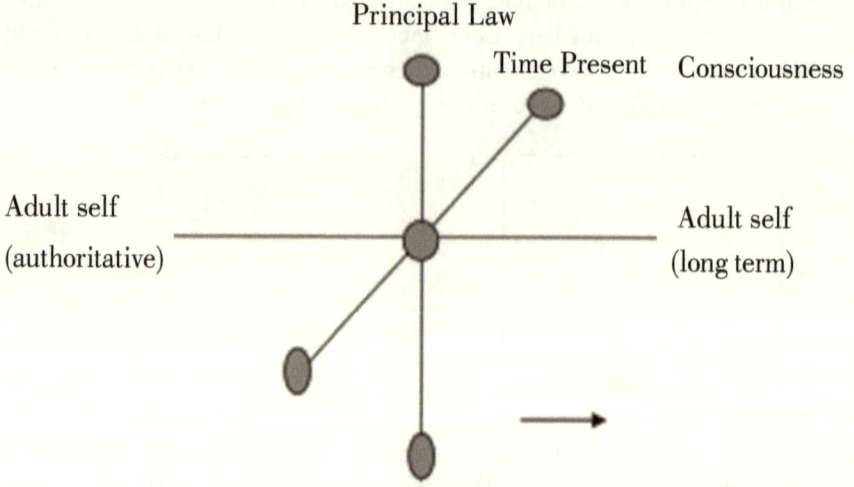

1:1   My use maybe Lazy or it may be self

guidance or both.

$$1+3 = 18$$

(use elsewhere or not 3 corner)

All corners of cube faces or the whole thing can be gathered to make purpose. I have mine set as an advanced still progressive always moving piece. My moving pieces come from within, not so much as what guides as external authority unless wise councel.

Science could use a piece or two for dominant genes.

Stress relief

Physiology.

www.ingramcontent.com/pod-product-compliance
Lightning Source LLC
Chambersburg PA
CBHW030812180526
45163CB00003B/1247